T0305883

Industry 5.0

Industry 5.0

The Future of the Industrial Economy

Uthayan Elangovan

CRC Press
Taylor & Francis Group
Boca Raton London New York

CRC Press is an imprint of the
Taylor & Francis Group, an **informa** business

First edition published 2022
by CRC Press
6000 Broken Sound Parkway NW, Suite 300, Boca Raton, FL 33487-2742

and by CRC Press
2 Park Square, Milton Park, Abingdon, Oxon, OX14 4RN

© 2022 Uthayan Elangovan

CRC Press is an imprint of Taylor & Francis Group, LLC

Reasonable efforts have been made to publish reliable data and information, but the author and publisher cannot assume responsibility for the validity of all materials or the consequences of their use. The authors and publishers have attempted to trace the copyright holders of all material reproduced in this publication and apologize to copyright holders if permission to publish in this form has not been obtained. If any copyright material has not been acknowledged please write and let us know so we may rectify in any future reprint.

Except as permitted under U.S. Copyright Law, no part of this book may be reprinted, reproduced, transmitted, or utilized in any form by any electronic, mechanical, or other means, now known or hereafter invented, including photocopying, microfilming, and recording, or in any information storage or retrieval system, without written permission from the publishers.

For permission to photocopy or use material electronically from this work, access www.copyright. com or contact the Copyright Clearance Center, Inc. (CCC), 222 Rosewood Drive, Danvers, MA 01923, 978-750-8400. For works that are not available on CCC please contact mpkbookspermissions@ tandf.co.uk

Trademark notice: Product or corporate names may be trademarks or registered trademarks and are used only for identification and explanation without intent to infringe.

Library of Congress Cataloging-in-Publication Data
Names: Elangovan, Uthayan, author.
Title: Industry 5.0 : the future of the industrial economy / Uthayan Elangovan.
Description: First edition. | Boca Raton : CRC Press, 2022. |
Includes bibliographical references and index.
Identifiers: LCCN 2021028869 (print) | LCCN 20210288702 (ebook) |
ISBN 9781032041278 (hardback) | ISBN 9781032041285 (paperback) | ISBN 9781003190677 (ebook)
Subjects: LCSH: Industry 4.0. | Society 5.0. | Automation. | Robots, Industrial. | Human-computer interaction. | Industrial engineering. | Internet of things–Industrial applications. | Artificial intelligence–Industrial applications. | Manufacturing industries–Technological innovations.
Classification: LCC T59.6 .E43 2022 (print) |
LCC T59.6 (ebook) | DDC 658.4/0380285574–dc23/eng/20211103
LC record available at https://lccn.loc.gov/2021028869
LC ebook record available at https://lccn.loc.gov/2021028870

ISBN: 978-1-032-04127-8 (hbk)
ISBN: 978-1-032-04128-5 (pbk)
ISBN: 978-1-003-19067-7 (ebk)

DOI: 10.1201/9781003190677

Typeset in Times
by codeMantra

Dedication

_To my parents, Elangovan Rajakani and Kanmani
Elangovan, who educated me on the qualities of self-control
over and above honors of education and learning_

Contents

Preface

The rate of modern technology creation and fostering certainly varies across different industrial sectors. Technology is frequently progressing, and production must develop with it to stay competitive. While Industry 4.0 is still the primary transformation for many manufacturing leaders, it's crucial to look toward the future. In the industrial sector, the Internet of Things represents an effective modern technology driving a lot of the modifications since the introduction of the Industrial Internet of Things, providing significant benefits to manufacturers from different industries. This publication endeavors to provide a glimpse of how small to medium enterprises and original equipment manufacturers can best leverage, increasing the process effectivity, operational effectivity, reducing unskilled workforce utilizing Industry 3.0 to Industry 4.0 through Industry 5.0, will definitely lead to the manufacturing of greater worth tasks than ever before, driving optimum outcomes from human-to-machine interactions.

I have always had a passion for advancement in product lifecycle management and computer-integrated manufacturing. Being a PLM and IIoT business consultant, I've had opportunities to assist manufacturing enterprises to enhance their business processes, finding methods to fix issues and help them start off their transformation journey. I feel obliged to share my knowledge along with my experience. I hope that the information and expertise offered here will awaken business leaders, product design and development professionals, manufacturers, industrial automation professionals, IT professionals, consultants, and academies to come to the realization that to improve engineering process, they need production effectivity, quality, and zero waste manufacturing ecosystem.

I wish you enjoy your reading.

Acknowledgments

I would certainly like to express my gratitude to several individuals who saw me through this book; to all those who offered assistance, talked things over. Thanks to Cindy Renee Carelli – Executive Editor, Erin Harris – Senior Editorial Assistant, my publisher CRC Press/Taylor & Francis Group: without you, this book would certainly never find its place in the digital world, and to a lot of individuals throughout this global village.

I thank Joel Stein for revealing the course to authoring this book.

I would like to thank my friends – Subject Matter Experts, who took part in the design thinking process – S. Palanivel, E. Srinivas Phani Chandra, A. Babu, K. Manikannan, V. Venkataramanan, V. Bhuvaneswaran, D. Gopinath, K. Gopinath, R. Selvaraj, E. Kamalanathan, N. Ganesh, S. Rajaprakash, P. Saravanan, A. Kalidhas, and P. Baskar.

I express my love and gratitude to my parents, my wife – Saranya Uthayan, my son – U. Neelmadhav, my professors, my good friends, associates in the business, and all my well-wishers, without whom this book would not have been possible.

Author

Uthayan Elangovan has 17 years of dynamic experience, ranging from product life-cycle management (PLM) to Industrial Internet of Things (IIoT) consulting for an assortment of businesses, including automotive, electrical, medical, industrial, and electronics enterprises. He helps and leads PLM, IIoT usage, and subventures, and with cutting-edge collaboration tools and techniques, he gives consultations to worldwide clients. Energetic about PLM, IIoT, and its effect on product development guaranteeing PLM, IIoT system meets client deliverables while supporting business processes. His interest in making technological advancements in automation influenced him to write his first book *Smart Automation to Smart Manufacturing – Industrial Internet of Things*, which was named as one of the Best Manufacturing Automation books of all time by Book Authority. His passion for PLM and IIoT influenced him to author his second publication *Product Lifecycle Management (PLM): A Digital Journey Using Industrial Internet of Things (IIoT)*, which was named as one of the Best Industrial Management books of all time, New Industrial Management books to be read in 2021 and New Product Design books to be read in 2021 by Book Authority.

He earned a bachelor's degree in mechanical engineering from Kongu Engineering College and a master's degree in computer-integrated manufacturing from PSG College of Technology. He currently resides in Tamil Nadu, India, and is a consultant for PLM and IIoT, providing business and education consulting through his firm – Neel SMARTEC Consulting.

List of Figures

Abbreviations

ADAS	Advanced Driver Assistance System
AGV	Automated Guided Vehicle
AI	Artificial Intelligence
AM	Additive Manufacturing
AMS	Aerospace Materials Specifications
APC	Advanced Process Control
APQP	Advanced Product Quality Planning
AR	Augmented Reality
ASPICE	Automotive Software Performance Improvement and Capability dEtermination
BOM	Bill of Material
BPA	Bisphenol A
BPM	Business Process Management
CAD	Computer-Aided Design / Drafting
CAE	Computer-Aided Engineering
CAM	Computer-Aided Manufacturing
CAPA	Corrective Action Preventive Action
CAx	Computer-Aided technologies
CFD	Computational Fluid Dynamics
CFT	Cross-Functional Team
CIM	Computer-Integrated Manufacturing
CNC	Computer Numerical Control
CRM	Customer Relationship Management
CTQ	Critical to Quality
CQI	Continuous Quality Improvement
DL	Deep Learning
DCS	Distributed Control System
DFA	Design for Assembly
DFF	Design for Fabrication
DFE	Design for Environment
DFM	Design for Manufacturing/Design for Manufacturability
DFMEA	Design Failure Mode and Effect Analysis
DFR	Design for Reliability
DFT	Design for Testing
DFSC	Design for Supply Chain
DFSS	Design for Six Sigma
DFx	Design for Excellence
DPA	Digital Process Automation
DMAIC	Define, Measure, Analyze, Improve, and Control
DMT	Defect Mapping Tool
DOE	Design of Experiments
DRC	Design Rule Checks

EaaS	Energy-as-a-Service
eBOM	Engineering Bill of Material
ECAD	Electronic Computer-Aided Design
EDA	Electronic Design Automation
EMS	Electronic Manufacturing Service
Ems	Environment Management System
ERP	Enterprise Resource Planning
ESD	Electrostatic Discharge
ESG	Environmental, Social, and Corporate Governance
FMEA	Failure Mode and Effects Analysis
FEA	Finite Element Analysis
FEM	Finite Element Method
FDA	Food and Drug Administration
GRN	Goods Receipt Note
GPU	Ground Power Units
GSE	Ground Support Equipment
HMI	Human–Machine Interface
IATF	International Automotive Task Force
ICS	Industrial Control System
ICT	Information and Communication Technology
IDOV	Identify, Design, Optimize, and Verify
IEC	International Electrotechnical Commission
IIoT	Industrial Internet of Things
IoT	Internet of Things
IPA	Intelligent Process Automation
IPC	Institute of Printed Circuits
IR	Infra-Red
ISO	International Organization for Standardization
IT	Information Technology
JIT	Just-In-Time
KPI	Key Performance Indicator
M2M	Machine 2 Machine
ML	Machine Learning
MES	Manufacturing Execution System
MSA	Measurement System Analysis
MDM	Medical Device Manufacturer
MRO	Maintenance, Repair, and Overhaul
MRP	Material Requirements Planning
MSD	Moisture-Sensitive Device
MVDA	Multivariate data analysis
MVP	Minimum Viable Product
NC	Numerically Controlled
NLP	Natural Language Processing
NPD	New Product Development
NPI	New Product Introduction
OEE	Overall Equipment Effectiveness

ODM	Original Design Manufacturer
OEM	Original Equipment Manufacturer
OCR	Optical Character Recognition
PCB	Printed Circuit Board
PCA	Printed Circuit Assembly
PCBA	Printed Circuit Board Assembly
PCA	Process Control Automation
PCS	Process Control System
PDM	Product Data Management
PEEK	PolyEther Ether Ketone
PESTLE	Political, Economic, Social, Technological, Legal, and Environmental factors
PLC	Programmable Logic Controller
PLM	Product Lifecycle Management
PMEA	Process Failure Mode and Effects Analysis
PPAP	Production Part Approval Process
PVC	Polyvinyl Chloride
QFD	Quality Function Deployment
QMS	Quality Management System
RCA	Root-Cause Analysis
RF	Radio-Frequency
RFID	Radio-Frequency Identification
ROI	Return of Investment
ROV	Return of Value
RPA	Robotic Process Automation
RPN	Risk Priority Number
SCADA	Supervisory Control and Data Acquisition
SCARA	Selective Compliance Assembly Robot Arm
SCM	Supply Chain Management
SME	Small to Medium Enterprise
SMT	Surface Mount Technology
Solar PV	Solar PhotoVoltaic
SPC	Statistical Process Control
SWOT	Strengths, Weaknesses, Opportunities, and Threats
THT	Through Hole Technology
TPS	Toyota Production System
TCP/IP	Transmission Control Protocol/Internet Protocol
VPVC	Unplasticized Polyvinyl Chloride
VSM	Value Stream Mapping
VR	Virtual Reality
WCM	World Class Manufacturing
WEEE	Waste Electrical and Electronic Equipment

1 Industrial Transformation

Manufacturing industries around the global village are on the threshold of great opportunities that promise extraordinary development and transformation of their business through smart products and smart manufacturing, enabled by cutting-edge technological innovation. Industrial sectors sell their products thorough complex processes such as research, design, development, manufacturing and service. Every product manufacturing segment has unique challenges that cannot be tackled by a one solution that fits all requirements. Manufacturing enterprises perennially encourage the development of science and technology and adopt a variety of approaches to transform their businesses, thereby constantly seeking new ways to upgrade and distinguish themselves from their competitors.

Digitalization has heralded a new paradigm in manufacturing, where manufacturing facilities are transformed to be extra modern and advanced. Consequently, this arouses concerns in the minds of business tycoons: will the emerging technologies take control of the manufacturing production line of futuristic factories? In a world of burgeoning modern technology, many manufacturers stand to gain much from automation, if the circumstances are exploited right. Taking automation to the next level can be a huge advantage for the manufacturing industry. Advanced automation can help reduce a holdup, reduce production expenses and enhance product quality.

Industrial sectors are reshaping their competitive landscape and steering in to a new era of growth, change and economic opportunity. Every organization requires their employees and machinery to do their jobs with greater efficacy and proficiency while managing operations, designing products as well as establishing intellectual property throughout the globe. The ultimate objective of industrial transformation is to achieve a better quality of product and service for the customer. Current business systems, including computer-integrated manufacturing (CIM), product lifecycle management (PLM), enterprise resource planning (ERP), manufacturing execution systems (MES), programmable logic control (PLC) and supervisory control and data acquisition (SCADA) along with Industrial Internet of Things (IIoT), are now being utilized to ensure that a superior user experience, quick time to value, integration of information and easy access from anywhere across the globe are realized. Innovation is making an impact on every stage action from product design to manufacturing.

Today, manufacturing industries are developing techniques for combining new innovations to improve their efficacy and performance, the leading concept behind Industry 4.0. It is essential to closely assess the elements of the business, from client connections to reshoring options and likely a lot more. Robotics has emerged to become the mainstay in production, and, Industry 4.0 innovations offer greater versatility in manufacturing processes. Manufacturers can also introduce new automation and artificial intelligence-assisted effectiveness to their enterprises. Heralding the next industrial transformation calls for the adoption, standardization and execution of new technologies, which requires its very own framework as well as advancements.

DOI: 10.1201/9781003190677-1

BUSINESS TRANSFORMATION

Business transformation is a strategic initiative on every business leader's campaign to remain competitive, which consists of workers, processes, as well as innovation to achieve measurable enhancements in effectiveness, performance and complete customer satisfaction. Organizations that continuously adapt are driven by a keen vision to redesign their future via transformation. An improvement is a major change in an organization's abilities and identity to ensure that it can deliver valuable outcomes, pertinent to its objective, which it could not accomplish previously. Business transformation is more defined by a high level of passion of the organization as a substantial space that should be linked in between the current and future enterprise path. It represents an essential enhancement in the present business operations. A robust commitment to value expansion is an effective directive for identifying the efforts that will certainly have the best influence on an enterprise transformation road map and also for understanding its prospective worth for investors.

> Among the successful business change instances is Apple – from being a producer of computer systems, Apple has slowly taken place to customer devices. Experts say the shift has been smooth. After the launch of iPod, Apple changed from being a hardware and software supplier, to the domain of customer electronic devices. With the launch of iTunes Music store, Apple became a media business.
>
> *(Gupta and Perepu 2006)*

Service improvement needs to consistently be a step in the right direction for a thriving business. Because of this, business transformations need to aim at making inroads in to entering a brand-new section of the marketplace, adding industrial value to the business, improving the efficacy of the manufacturing processes and making best use

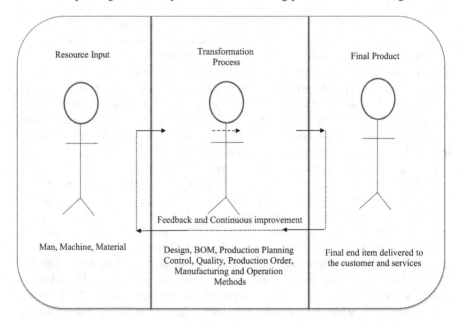

FIGURE 1.1   Simple transformation process.

of the available resources. Business advancement aspects differ for every manufacturing enterprise. This is because every enterprise has their own strength to leverage and difficulties to deal with. The path toward business transformation is never easy, as it is fraught with challenges. Irrespective of the nature as well as the objective of the transformation, all enterprises can anticipate significant resistance to change. For a successful tranformation, the management must dare to take risks and must be steadfast and meticulous in its execution. The success of a business transformation squarely built on the ability of the enterprise to adapt to change in strategies often determined by market change, disruptive needs and tactical direction. The ability of the management to overcome these obstacles is one of these the crucial success variables.

KEY ELEMENTS OF BUSINESS TRANSFORMATION

Manufacturing industries need to direct their attention from mere survival to seeking new methods to grow. Considering globalization and the fast pace of today's business, there are no quick fixes for simplifying business intricacies. Globalization has made it hard to keep organizational frameworks simple. Today's multinational businesses have thousands of employees, numerous organization companions and substantial operations spread across the globe. Following an appropriate enterprise organization structure and also operating model is an ongoing battle.

Few questions that can help manufacturing industries to understand the need for business transformation are as follows:

1. How pleased are the clients with your product and service?
2. What are the different ways to enhance client experience?
3. How to prosper in the current smart and connected competitive world?
4. How would certainly financial investments in technology improve the experience?
5. How can success be determined?

Manufacturing industries cannot transform successfully unless their individuals within the enterprise transform; majority of the transforming initiatives fail because enterprise overemphasizes the tangible side of the transformation. Business transformations within an enterprise impact the monitoring of operational settings, interfere with the cultural norms, modify service procedures and capitalize on new modern technologies. Some examples of buissness transformations in industrial sectors are as follows: organization transformation, technology transformation, business process transformation and industrial transformation.

ORGANIZATION TRANSFORMATION

Organization transformation is a basic, enterprise-wide change impacting how a company is run while focusing on augmenting its efficiency and proficiency. Organization transformation is a term that refers collectively to activities such as reengineering, revamping and redefining organization systems, and it happens in response to rapidly changing demands and the compulsive need to improve the enterprise's efficiency along with sustainability. It shows the measures adopted by the business leaders to steer the business successfully into the future and to achieve the desired result. However, if the company perceives delays in its quarterly reports, it might have a much more substantial issue on its hands. As every business experiences cycles of development along with change, this is an oppurtunity to analyze the performance of the company and prepare a strategic plan for its future. What is required is an alternative procedure that companies can utilize to help them incorporate as well as implement changes throughout the organization.

> Google achieved organization transformation by developing higher division. Research and development division dealt with such a variety of projects that it was ending up being tough for management executives to concentrate on innovation. Tactical solution devised is splitting right into several business entity, each of them with a slim focus, responding to the new parent firm Alphabet.
>
> *Alphabet Inc. (2017)*

Maintaining a keen eye on both the problems will provide an insight into whether or not any organization transformation is needed. Change is usually driven by C-level executives who are in charge of process of the organization. It is important for the success of any transformation program that the organization rightly identifies the reality and is prepared to adopt the required procedures without losing focus as the organization transformation initiative is implemented. Transforming an organization requires the ability to be agile, receptive to market trends and technology, whenever essential. These adjustments are lasting only when they affect the end users to alter their actions and influence supervisors to adapt and approve brand-new concerns. Organization transformation is more likely to do well when the organization agrees to accept the change and when the scheduled modification is integrated well with existing business control systems and also culture. Transformational changes call for

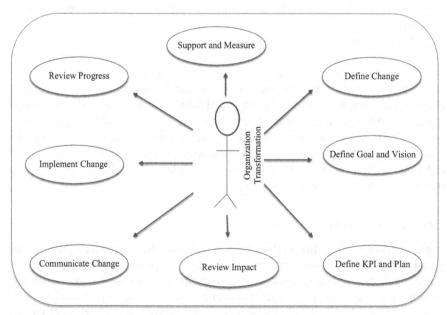

FIGURE 1.2   Organization transformation.

considerable advancement besides learning. Business participants need to discover exactly how to enact the new ways and carry out new approaches. Regardless of the level of the organizational transformation, it is critical that the organizational effect and threat analyses are carried out to permit C-level executives to recognize the resources necessary to efficiently execute the modification effort and to establish the impact of the modification on the organization.

BUSINESS PROCESS TRANSFORMATION

Business process transformation is the essential rethinking of a process within an enterprise. This focusses on the end-to-end placement of the main purposes, measures, information, metrics and innovation in accordance with the strategic goals and also the tactical needs of the business, delivering a significant, calculated increase in client value. It entails an assessment of the actions called for to attain a specific objective in an effort to remove replicate process tasks. Identifying a process will aid in saving time, speed up the return on investment and return on value and save sources. It is essential to recognize the best alternatives in order to pick the best technology and application plan to sustain both business process transformation needs and strategic goals. Primarily, business process transformation is driven by market needs and entails automating as many procedures as possible.

> "Siemens Vision 2020, which outlined an organizational overhaul, restructuring, and also calculated shift from energy and also commercial manufacturing to digitalization." "Philips Split its lights core from its medical care growth company, changing itself into a health care modern technology firm."
>
> *(Anthony et al., 2019)*

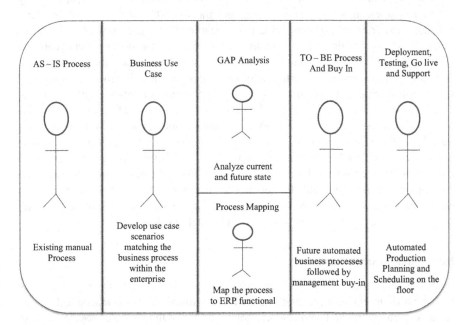

FIGURE 1.3 Business process transformation – ERP implementation.

Business process transformation is a deliberate and organized extension of the transformation journey that garners a substantial return on investment leading to breakthroughs in an organization's efficiency outcomes. Whatever the target or nature of the business change, the goal is always to reinforce its relevance in the competitive market in order to guarantee its survival. Thus, business process transformation is a beneficial goal and a vital building block for any significant and calculated changes made within the enterprise. It is arguably a crucial action in every kind of service modification. During the COVID-19 pandemic, several transformational fads were set to speed up, swiftly. Many companies in different industrial sectors were under tremendous stress to quickly adapt their service designs in this radically altered market conditions in order to stay sucessful. The most significant risk to any effective business process transformation is aligning it with a suboptimal data method. Like organization transformation, business process transformation too requires cautious planning, clear objectives and confident administration.

TECHNOLOGY TRANSFORMATION

Technology transformation is a vital part of competitive business practices today in this smart connected world. Most of the industrial application systems used in the enterprise are attracted by new cutting-edge technological advancements by innovative companies with response to expectations of customers. These enterprises are consistently evolving their internal information technology ecosystem to minimize hazards and simultaneously boost business continuity. Industry 4.0 is currently sweeping the industrial economic scene in a manner similar to the impact of the mass media and communications on the industry over the past decade.

Innovation in every industrial sector supports the creation of all new, digitally enabled business models, while holding out the important assurance of boosting consumer experiences and enhancing the productivity of legacy process. The information and communications technology revolution is transforming conventional sectors, ensuing changes and big modifications in well-established ecosystems. Advanced innovations are vital to modern business, and, it is fair to claim that every big manufacturing sector needs to move toward this transformation to attain growth. Technologies transformed the method individuals functioned, but they did not fundamentally alter the way businesses ran. Certainly, technological innovation can itself be a driver for massive organizational changes such as the method by which employees interact with each other and the manner in which the business engages with customers, companions and also various other stakeholders engage. The COVID-19 pandemic has augmented the demand for driving technological transformation across businesses of all dimensions. Enterprises are welcoming remote jobs and swiftly customizing their daily procedures to match the new normal.

INDUSTRIAL TRANSFORMATION

Industrial sectors are continuously growing and transforming by seeking essentially new methods to enhance monetary assets besides functional performance, safety and security, high quality and competitive advantage. One of the main developments is the exchange of information between during various stages of the customer service. Industry leaders often challenge their internal groups to come up with the most effective and innovative ways to transform their businesses utilizing digital innovations, improving value chain processes and through collaborative work environment to serve the markets better. An effective industrial transformation requires not only innovation but also a shift in the perspectives of the people who eventually apply and also utilize the new processes.

TRANSFORMATION IN INDUSTRIAL MANUFACTURING

Technology is omnipresent and continues to transform numerous sectors particularly the production market. Business enterprises need to continually update on technical development, if the industries have to remain competitive, which implies that manufacturing services need to be aware of what is the best and also the most recent innovation. Innovative methods are adopted in the process of selecting the material and made use of in the manufacturing process; this has made the process less labor-intensive and much safer, consequently boosting the operation. The use of computers and smart devices each day has actually boosted the performance of the enterprise residences, offices and manufacturing facilities are being managed utilizing smart tools, and the industrial sector's transformation trajectory is progressive. The demand for smart products is surging, calling for new and ingenious manufacturing approaches, and numerous producers have actually stepped up to the next level of industrial transformation. Recognizing the nature of the change in the production sector will certainly aid to establish which methods are likely to prosper and which will not.

At the end of the eighteenth century, manufacturing industries steered away from the artisanal approach and moved toward machine-assisted work, and when electrical power was utilized in manufacturing facilities, it led to real mass production becoming a viable choice. Possibly, one of the most impactful changes has been the transition to automation. The introduction of computers in production became more and more noticeable, with digital systems being developed to oversee a whole assembly line. Presently, customers want things faster and also better, customized and distinct. Consequently, manufacturers have to not only find a method to maintain the demand for products but also find skilled workers to make these products.

The advancements in technology and the consequent growth of the industry have improved the status of the business and gained the trust of its customers. Scientific research and development along with computer simulation in product advancement as well as other locations have made a significant influence. Every new advancement has brought a change in the production process that has in fact transformed the way the industries function across different sectors. Generally, the growth of industrial transformation shows the complete understanding of how technological innovation improves the growth of manufacturing industry. The application of industrial robots along with artificial intelligence is on the rise and they are much more sophisticated and proficient at doing intricate jobs. Even the expense differential with human beings is narrowing, to the advantage of robotics.

Transformation from analogue, mechanical, along with digital innovation, describes the path to digital innovative technology. Bridging the gap between design and manufacturing through services and making use of big data, data analytics and machine learning transform to the fourth industrial revolution. Understanding the importance of data generated across the manufacturing enterprise opens up opportunities for the evolution of new business models that transition from being product centric to customer-focused service centric.

FUTURE INDUSTRIAL TRANSFORMATION

A great deal of the innovations that are in practice currently are facelifts of the fundamental ideas laid throughout the transformations that took place so far. The fifth industrial transformation, also known as Industry 5.0, is already becoming part of the industrial automation landscape. Industry 5.0 combines human creativity and workmanship with the rate, effectiveness and also consistency of robots. Moreover, it complents human beings by whetting their creative thinking. Industry 5.0 creates even higher-value tasks than Industry 4.0, due to the fact that human beings are reclaiming product design through manufacturing that calls for creativity.

> Robot guides, group control directed by artificial intelligence and also immersive virtual reality are among the technologies, allowed by the internet of things, set to delight fans at Tokyo's 2020 Olympics. Robots (Field Support Robot, Remote Location Communication Robot, Human Support Robot, Delivery Support Robot) established by the Toyota Motor corporation will certainly help spectators in a series of tasks, from carrying food and also various other items to showing people to their seats and also supplying information on occasions. The robotics also help functional real-life implementation aiding individuals, besides visitors in mobility devices.
>
> *(Olympics, 2019; Forbes, 2019)*

To strike a balance wherein the machine-human interaction can supply the highest possible benefits, where increasingly complex processes will certainly call for an ecosystem that is capable of handling the substantial amount of information generated and also provide human operators with a room that they can utilize to connect with shop floor machines with the development of digital twins. Industry 5.0 combines human creativity and robotic accuracy to engender a distinct option that will soon be in demand of the coming years. Both Industry 4.0 and Industry 5.0 have paved a road map that industries can/shall follow in order to sustain.

SUMMARY

Technology-driven transformation requires the appropriate organization culture and management executives to function appropriately. Modern technology alone is not enough to drive these transformations; business leaders need to engage with their workers to encourage understanding and adoption. Manufacturing industries that takes care of to foster the appropriate culture to incorporate these new technologies will be the ones with a competitive advantage, improving their existing business models, developing new possibilities, all the while retaining time-tested skills and simultaneously drawing in brand-new skills. Strategic investments continue to be vital for every manufacturing organization's ongoing development; even if different aggregating techniques in varied operations can be made complex, the process can aid manufacturers to see high returns in the increasingly competitive environment. This is really a future that provides value to the manufacturing. A key aspect in improving business performance is having the most efficient processes and the most effective people, focusing on a client's outcomes and using cutting-edge technology to identify areas for improvement to leverage engineering process effectiveness through manufacturing effectiveness across the different levels of the enterprises.

BIBLIOGRAPHY

Alphabet Inc. 2017. *Reorganizing Google* (Case Code: HROB185). Hyderabad: IBS Center for Management Research (ICMR).

Anthony, Scott D., Alasdair Trotter, Rob Bell and Evan I. Schwartz. 2019. *Transformation-20.* Boston, MA: Innosight.

Becker, J., M. Kugeler and M. Rosemann. 2010. *Process Management: A Guide for the Design of Business Processes.* Germany: Springer.

Bradford, M. and G. J. Gerard. "Using process mapping to reveal process redesign opportunities during ERP planning." *Journal of Emerging Technologies in Accounting* 12 (2015): 169–188.

Engel, A., T. R. Browning and Y. Reich. Designing products for adaptability: insights from four industrial cases. *Decision Sciences* 48, no. 5 (2017): 875–917.

Forbes. 2019. https://www.forbes.com/sites/stevemccaskill/2019/07/29/tokyo-2020-to-use-robots-for-a-more-efficient-and-accessible-olympics/.

Gupta, Vivek and Indu Perepu. 2006. *The Transformation of Apple's Business Model Case Study* (Case Code: BSTR212). Hyderabad: IBS Center for Management Research (ICMR).

Hammer, M. and J. Champy. 1993. *Reengineering the Corporation: A Manifesto for Business Revolution.* New York: Harper Business.

Harrington, H. J., D. R. Conner and N. F. Horney. 1999. *Project Change Management: Applying Change Management to Improvement Projects.* New York: McGraw-Hill Trade.

Kane, G., D. Palmer, A. Phillips, D. Kiron and N. Buckley. 2015. *Strategy, Not Technology, Drives Digital Transformation.* Texas, MIT Sloan Management Review and Delloite University Press.

Madison, D. 2005. *Mapping, Process Improvement, and Process Management: A Practical Guide to Enhancing Work and Information Flow.* Chico, CA: Paton Press.

Manganelli, R. L. and M. M. Klein. 1994. *The Reengineering Handbook: A Step-by-Step Guide to Business Transformation.* New York: AMACON.

Olympics. 2019. *New Robots Unveiled for Tokyo 2020 Games.* https://olympics.com/ioc/news/new-robots-unveiled-for-tokyo-2020-games.

2 Engineering and Manufacturing Transformation

Manufacturing organizations aim to integrate a varied set of functions such as quality control, supply management and so on to collaborate in a streamlined way. Having stated that, the primary focus on business enterprises is automating a variety of product development activities such as functional designs, procedures management, system simulations and recurring performance analysis of each step in the manufacturing of a part. The process of bringing information technology-driven automation in design and production tasks requires process automation in engineering through manufacturing. By reducing the time to carry out each activity, companies can achieve significant financial savings throughout the enterprise. The ideation behind automation is not recent; the concept of using automation has actually been in practice for years; however, it has become more popular and essential for certain industries in the last hundred years. Throughout Industry 1.0, Industry 2.0 and Industry 3.0 automation had been mainly implemented in industrial grounds; Industry 4.0 integrates industrial automation with information and communication technology. With industrial automation, the objective was clearly to boost the effectiveness of manufacturing customized products. Automation of a few crucial processes can enhance the efficiency of specific processes, which is one of the most persuading factors for organizations to take on process transformation.

From medical component to industrial component manufacturers and from automobile manufacturers to aerospace and defense sectors, finding innovative methods to reduce expenses and save time to market while consistently supplying high-quality products will be an essential element for all industrial sectors. The business enterprises are perpetually nurturing out-of-the-box thinking as well as revamping the business process and utilizing new methods to transform the conventional approach to design. CIM along with computer-aided engineering (CAE) innovation supports the collaborative processes required to make a substantial impact on the product design life cycle, enhancing functional performances besides lowering prices. PLM supports to handle a vibrant set of engineering files that maintain a history of design changes and ensure that new product development/new product introduction (NPD/NPI) teams are always working with the most recent updated product data. It sustains design effectiveness by giving a solitary resource of the right information, accessible in the best context. As rises in computer advancement in product design and development rise exponentially, advanced sensing unit technology, robots and artificial intelligence (AI) control systems along with various other technology developments pave the path to the future wherein smart manufacturing impact seems positioned for a remarkable change in many industrial sectors.

DOI: 10.1201/9781003190677-2

PROCESS AUTOMATION

Industries encounter several difficulties as globalization continues to decrease the earnings margin, but at the same time, they are required to produce quality products and services. Automating everyday tasks ensures procedures are carried out in a prompt manner, without missing out on the target date. Automation predominantly takes over all the labor-intensive tasks, making it possible for an organization to speed up operations substantially with few mistakes. With enhanced effectiveness comes increased capacity, making it easier to scale operations as the business grows. Process automation allows members of the NPD team to carry out even more innovative jobs that are more rewarding and satisfying; this contributes to the organization's success. It is essential to have a business strategy that combines more practices to simplify complex processes. Process automated workflow-enabled product development procedure can be used to enhance, standardize and shorten the development cycle.

Process automation is an essential function of digital transformation

Introduction of new product into the market is one of the basic strategies for any type of manufacturing industry. New products are introduced in the market annually as customers are looking out for more assortment of products. High competition forces business enterprises to reduce the cost of the new product development to comply with a quantifiable process to produce renewed market offerings. Development of process automation entails the assimilation of process, people and data along with software applications throughout the enterprise to automate process-oriented tasks. Enterprises that are not able to produce new products to the market can experience the repercussions. Integrated automated process can connect spaces and in addition break down silos to advertise partnership in between various groups and also parts of business, helping with cross-functional team involvement.

Process automation is appropriate for type of industrial sector, although each sector has distinct business policies that control how tasks are executed. Implementing process automation might differ for different industries, e.g., product information management comprises of product lifecycle management, enterprise resource planning, supply chain management, and customer relationship management. There is still dilemma in the minds of small to mid-size manufacturers on whether process automation is the the most essential thing that the enterprise needs right now. Financial investment in automation ought to belong to a broader area in the product design and manufacturing process. Plainly, this ought to be lined up to the enterprise strategy planning. Initiating partial automation is a much more practical objective for some enterprises, particularly small to medium enterprise (SME) manufacturers. They remain in a tough spot while considering how to take the right steps toward embracing a new technology along with the business processes.

ENGINEERING PROCESS AUTOMATION

Automation in product design and development is one possibility to sustain product designers in their everyday work. Design processes are complicated and include cross-functional teams that include sales, advertising and marketing, designing, manufacturing and quality control. Assimilation of customer needs and market demands is directed to product developers and designers. Manufacturers intend to bring a diverse collection of cross features to interact in a structured fashion. Process automation in designing easily accomplishes this for SME manufacturers, besides providing quicker reaction to market needs. Product design and process layouts have been in practice for years in different industrial sectors. Even a small initiative in automation can boost the efficacy of many individuals procedures, as it is one of the most convincing factors for an enterprise to adopt process automation. Engineering process automation allows product manufacturers at all levels to turn around propositions swiftly, design and manufacture effectively and cater more efficiently to consumer demands and generate a healthy, competitive and balanced revenue.

Product manufacturers utilize a series of engineering applications for process automation systems such as CAx which includes CIM consisting of computer-aided design (CAD) and computer-aided manufacturing (CAM) and CAE which have control over process automation. Integrated product database permits an enterprise to focus on its data management efforts along with lifecycle management. Product configuration practices can be superimposed on top of product data management features. CAx tools along with product data management (PDM) and PLM make it possible for engineering changes, assisted in with the product configuration management to be better created and also evaluated to define changes to be implemented in new product development and new product introduction process.

Engineering process automation aids manufacturers in enhancing the design, analysis and manufacture of products. The journey of engineering process automation starts with preserving the particular of the product and process design details electronically, which will reduce paper-based representations. This reduces hand-drawn drawings and the storage of papers; it also minimize the time to access the most updated version of a part along with its associated drawings and reduce errors. Making use of these technologies and combination concepts effectively improves the communication regarding the product and process design within the design function and across the extended enterprise and clients.

MANUFACTURING PROCESS AUTOMATION

Manufacturers are always looking out for ways to reduce the production operation cost as well as speed up the manufacturing process without compromising on the end product quality. It has motivated manufacturers throughout all industrial sectors of all segments to automate business workflows, along with integrating other enterprise systems. Manufacturing processes automation is one such critical component of nearly all product manufacturers across the globe. It provisions the information needed to take care of the shop floor operation as a whole. Products can be tracked from ideation, creation, right to the customer.

Business enterprises attain better productivity by installing manufacturing automation in all operations and interaction leading up to manufacturing. Manufacturing process automation is the topmost priority of most manufacturers. It aids to carry out operations along with tasks such as processing, production setup, examination, inventory management and production preparation. Manufacturers are additionally integrating real-time dynamic surveillance, quality management, ERP though MES to acquire greater precision, range and faster time to market. The main function is to reduce the manufacturing lead time, reduce wastage and ultimately boost the productivity qualitatively and quantitatively.

Some of the manufacturing process automation tools available for manufacturers are as follows:

- Distributed control system
- Programmable logic controller
- Supervisory control and data acquisition
- Human–machine interface
- Artificial neural network

These are used to integrate the flow of inputs from sensing units and events with the passage of results to actuators, instrumentation, movement control and robots.

Business Process Automation

Constant changes in customer needs as per the latest market trends in the industry, quick decision-making and delivering quality customer service along with running the organization more efficiently can be achieved with the help of business process automation (BPA). The core principles behind the BPA function are orchestration, integration and automated execution. With automated processes in position, enterprises save time and make sure that the finest techniques are applied to improve general functional effectiveness. In the age of Internet-enabled manufacturing, process automation has risen to become the most sought after innovation that delivers the finest service to both internal and external customers. BPA has become the norm of process standardization for process quality and constant improvement, with a combination of enterprise systems such as PLM, ERP and MES. The arrival of Industry 4.0 technology Industrial Internet of Things (IIoT) connected machines to the digital environment; it enabled the movement of information over a network with no human communication.

Process automation and management is the foundational structure of product lifecycle management. It automates and also accelerates repeatable production business processes such as product design modification demands, authorizations, engineering and manufacturing change orders, along with automated workflows. Process monitoring brings a methodical approach to the tasks executed by both the NPD team and enterprise system. Most of the recurrent approval or review processes with a well-known reasoning in the product maturity model can be enhanced with PLM-based process management. Taking full advantage of the extent of the product growth process, PLM can be leveraged besides being recognized with the help of BPA. It enhances the new product development process, elevating their capacity to use product details to make the most effective general selections around which component to make, and how to build it.

One of the most vital parts of organization process automation is that it gets rid of silos throughout the manufacturing enterprise. Adopting process automation gives us the capability to access and also interpret service information throughout the business from the production line procedures to the business besides throughout the supply chain. Yet, at times, manual operations and disconnected systems can impede the ideal information from reaching the correct area at the correct time. Manufacturing enterprises make the best use of financial investments in existing modern technology and also attain higher degrees of interoperability than feasible with manual processes. Data access is less complicated and search details can be recorded over time, as it is electronically captured and stored in the cloud. Additionally, BPA is a substantial property to conformity procedures enforcing discipline, simplicity, and effectiveness right into the process.

ROBOTIC PROCESS AUTOMATION

Industrial manufacturing sector has predominantly seen automation equipped by physical robots that put together, examine package the products and aid in streamlining the assembly line. In contrast, robotic process automation (RPA) is a type of software robot that imitates the activity of a human being in performing a task within a process. It can do recurrent tasks more quickly, accurately and consistently than people. RPA allows the manufacturing enterprises to focus additionally on product technology and core strengths rather than the daily repetitive jobs that are essential but mundane in nature. RPA can be thought of as an electronic spine linking all applications. Manufacturing companies are incorporating digitization into their processes to enhance development, high quality and also performance. With RPA, manufacturers can update their production procedures and also service functions, accomplishing high productivity and labor reduction.

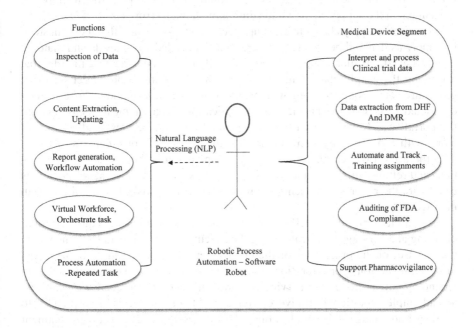

FIGURE 2.1   RPA role in medical device segments.

RPA technology can considerably assist in automating more applications such as supply chain process, product design, advancement and assistance to provide even more worth in the long term. RPA in manufacturing has absolutely become a vital enabler for procedure automation. The primary action initiated in the production service where RPA is incorporated is the selection of one of the most effective organization procedures to attain automation. The RPA automates processes such as extracting and updating details from numerous enterprise applications such as PLM, ERP, SCM and logistics companies that are incredibly taxing and also prone to errors.

RPA is not a replacement for the underlying business applications; instead, it simply automates the current manual jobs of the employees. One of the essential benefits of RPA is that the devices do not change the extant systems or infrastructure. It is a crucial tool in automating obstacles within the manufacturing enterprises pushing to grow toward their digital transformation journey. Technologists around the world are trying to upgrade automation efforts by infusing RPA with cognitive innovations in automating higher-order jobs.

Practice of RPA in Supply Chain

At the core of supply chain exists inventory control tracking. Distributors constantly require information regarding their supply levels to ensure they have sufficient products besides spares to meet customer requirements. RPA makes supply tracking simple by preserving details on supply levels, thereby informing the supply chain managers when the product supply levels decrease, and additionally reordering items right away. Business enterprises such as retail and production across markets have trusted applications such as PLM, ERP, MES, RFID, CRM and SCM. Of these, supply chain tracking uses RPA to automate common, low-value jobs, which simplifies procedures in the supply chain while eliminating human error.

RPA allows supply chains to scale up quicker to make certain that they can stock up supply as demand boosts. RPA figures out the needed stock levels and matches them versus the available supply. It automatically initiates the increase in order that travels up the supply chain without relying on human initiation. Order handling procedure gets automated along with payment such that information can be straight consumed right into the business data application. Negotiation happens through the portal, which can refine the recommended amount, besides send out e-mail along with SMS message confirmation for the positioning of order. As of today, with the arrival on AI, numerous supply companies trust bots to automate situations and refining them too, and by automating this back-office work, organizations can assure their workers concentrate on excellent quality tasks that require human intelligence.

The implementation of RPA is not as easy as it sounds; it requires proper planning at each stage and strategizing the execution procedure and acclimatizing the mindset of the employees of the enterprise; all these would certainly assist to develop the tone for alteration. If an enterprise aims to work with a digital technological improvement service provider, it needs to proceed with the end to end implementation initiative of Industry 4.0 in enterprise supply chain to avail full advantage along with return of value (ROV) and return of investment (ROI).

DIGITAL PROCESS AUTOMATION

Digital process automation (DPA) is advanced, digitized as well as automated, which is similar to typical business process management (BPM). DPA is a holistic strategy toward automation, which collaborates various modern technologies to transform the entire organization. DPA improves on the conventional BPM to aid organizations to better optimize end-to-end processes and to support enterprise-wide transformation. DPA assumes that business processes have currently been digitalized and focuses on optimizing the existing workflows to enhance the customer and end-user experience rather than incurring. Organizations that have experience with BPM ought to have the ability to easily execute DPA. DPA looks to automate not just business functions but also data monitoring within the business to make information available in real time. Essentially, DPA assists companies to transform into digital, data-driven businesses via automation and encourages employees to make data-driven decisions. Process automation can take care of numerous jobs as well as interactions, bringing together different information systems to supply manufacturers much less variation between processes, improved data honesty and storage space, much more effective and open to the shop floor process.

> Straightforward day today issue that DPA can address immediately goes out stock and supply replenishment of resources allowing the smooth circulation of details and automation of little work to let the internal team and customers understand when basic materials are back in supply. Implementing DPA is simply one of among the most essential improvements making enterprises can make throughout the transfer.
>
> *(Watts, 2020; Bizagi, n.d.)*

In this fast-paced smart connected challenging digital transforming world, manufacturers need to focus on innovation and performance to remain ahead of the competition by swiftly automating processes to supply excellence. DPA has a massive scope of advancement with IIoT, artificial intelligence (AI) and machine learning (ML) along with intelligent robots. Data play a big role as they make use of cloud-based options and help in evaluating the collected information to fulfill consumer demands and also improve the production process. Generating the silos of details from each other will certainly improve product design, promote manufacturing renovation, speed up delivery and boost end customer experience.

INTELLIGENT PROCESS AUTOMATION

The introduction of digital transformation across manufacturing industries integrates the internal business processes, thereby automating all interactions with the customer effortlessly. Business leaders adopt automation strategies to find the ideal process that result in minimizing the waste, improves cost efficiency and guarantees a better consumer experience. Consumers have extremely high expectations these days; hence, most manufacturing enterprises operate at an increased rate. Product manufacturers are expected to be a lot quicker in their response. Utilizing dynamic data to generate the most relevant results has become mandatory. Therefore, welcoming automation as become a vital method of functioning in today's world. Industry leaders are beginning to head in the direction of new innovations called intelligent process automation (IPA).

The idea of automation in the digital globe is gaining ground and technology is advancing with machines becoming more adept in performing human tasks. Innovations in digital modern technologies, accessibility of sensing units and enhanced computing power as well as storage have resulted in increasing its reach far in the worldwide technological landscape. It is the innovation that makes it possible for organizations to automate processes that involve structured, semi-structured and disorganized file systems consisting of records, text, photographs, videos, etc. It efficiently performs activities carried out by people. Some core modern technologies that develop the structure for intelligent process automation are AI, natural language processing (NLP), optical character recognition (OCR), smart workflows and RPA.

Manufacturers can make use of predictive analytics to fix brand-new issues and also promote product engineering. Machine learning models are being used to anticipate just how to produce specific physical products and also lessen anomalies in intricate material properties. The implementation of IPA helps the manufacturers to save time that would otherwise be invested doing computations to notice anomalies prior to a product enters quality assurance. IPA has set off industrial transformation in both Industry 4.0 and Industry 5.0, radically changing the processes and settings that depend on cyber-physical and cognitive systems.

IPA is the next level in the development of process automation and integrates artificial intelligence capabilities typically with process automation. It is developed to help employees by taking over repeated, regular jobs. It mimics human activities, without the demand for human intervention.

NECESSITY FOR PROCESS AUTOMATION

The most important variable that determines profit besides client satisfaction is process automation. Automating process in the business enterprise can be profitable, as it reduces expenses and improves the efficacy of the tasks.. Automation conserves money and time; however, what is right for one enterprise may not be right for an other. Moreover, the tasks that can be automated and the extent of automation are all factors that require consideration.. Determining the appropriate automation required can be an obstacle, yet it should not prevent the manufacturing organization from beginning to apply automation within the processes followed.

Developing strategies as well as formulating the real road map for automating the process is the first and foremost step to do. Nonetheless, strategy throughout the organization has its share of obstacles. Nevertheless, there are many business enterprises utilizing some kind of process automation within their business.

> I can think of an easy and most widely seen examples is that of even a small enterprise do follow process automation in the business is "Automatic reply" mail from the corporate email account to the clients.
>
> *(Luther, 2020)*

Today's fast-growing industrial economic pressures have considerably impacted the effectivity and efficiency of the industries encouraging them to adopt the process of automation. With leaner operating budgets, organizations can no longer manage to waste money on time-consuming and also tiresome jobs. The key reason for automation to have been welcomed so widely is its potential to yield higher results and increased productivity, aiding both cross-functional team members and other services. Innovation in robotics, industrial vision, IIoT, AI and cobots have opened lots of new capacities, allowing automation to be used not only in mass production processes but also in high-mix low-volume production environments.

Process automation levers numerous tasks, exchange of information, bridges the gaps as well as paves the path toward business process transparency across the enterprise. It involves running design data management, bill of material, vendor development and production planning control workflows by compiling information from CAD, PDM, PLM, ERP and MES.

As modern technology continues to become more advanced each day, manufacturers have the ability to run the business enterprise with leaner operating budgets, and designing, evaluation and other procedures are being replaced by smart process and smart machines that eclipse the abilities of human beings. Business decision-makers desire their shop floor machines to deliver the highest possible result with as little production expense as feasible. Process automation is very effective as it helps to increase design and manufacturing effectiveness, product and process quality, by reducing human assistance and the risk of errors. A majority of the manufacturing industries started to implement Industry 4.0 to process automation, which is considered to be a good indicator in the direction of digital transformation. Right automation technology coupled with the right skilled workers will position the buisness enterprise for success; the enterprise ought to begin the digitalization process in a stepwise method to meet the organization goal toward transforming in to a digital enterprise.

PROCESS TRANSFORMATION

Transformation is essential for many functions within a manufacturing enterprise. The journey starts with an effort to address a specific challenge, and soon after, the firm recognizes that they can implement the transformation and benefit if expanded throughout the enterprise. Initiating process transformation is one of the most efficient means to increase product quality along with operational performance. The leading concern of most manufacturing sectors is to improve quality. Process transformation in an industrial sector deals with quicker responses to market demands, and it goes to the core of the business process transforming it to take advantage of the digital capabilities.

As the rate of industry accelerates development becomes vital; enterprises are required to upgrade their strategies and processes to get new products off the attracting board as well out into the market rapidly. Yet, individuals who carry out the job of creating new products and designers are usually the most resistant to change. Most importantly, quality control approach includes decreasing risks, boosting training, and developing much better processes, besides making the work environment much safer and cleaner for everyone who spends time within it. The introduction of high quality transformation commences by getting rid of manual manufacturing process and also transitioning to touchless manufacturing. Incorporating business processes transformation into the decision-making process not only aids organizations get more worth out of their financial investment in innovation but also addresses issues concerning employees, helps them acclimatize to the transformation better. Overall, organizations are required to comprehend that utilizing modern technology to transform their businesses is fantastic, but the trick to succeed hinges on learning what ails the business using modern technology and correcting those issues before discovering novel uses of any kind of new age innovation.

As modern technology continues to progress exponentially, it has become imperative for firms to transform their process in order to secure a competitive advantage. Business leaders in the manufacturing industry have actually been traditionally reluctant to transform and introduce innovation due to the prices connected with making those changes. Leverage process, knowledge, skills along with modern technology to recognize its complete influence of process transformation, which needs to be part of the initial discussion ensuring future road map to success.

Transformation can just begin with a personal change of employees and by enhancing the process within the enterprise. Manufacturing ventures, be it small, medium or big, all should be ready to prosper in a globe where overhead prices, production operation cost along with resource cost are affordable. Furthermore, the organization will certainly need to specify the manufacturing objectives and set a clear strategy for transformation. Process transformation succeeds only after substantial thought has gone into identifying the potential improvement areas and enhancing a strategy where technology enhances service, increases profits and also reduces costs. Comprehending the different stages of the production venture maturity can assist process automation with process transformation in the manufacturing industry appropriately. It will certainly transform the way service is performed in every field of the industrial economy.

PROCESS AUTOMATION TO PROCESS TRANSFORMATION

Process automation involves using modern technology to make processes run themselves, making those processes much more effective and efficient with analytic reporting capabilities. The implementation of automation has reduced the time and has enabled supplier interactions to take advantage of customer interactions with quick, customized reactions at scale. In other words, process automation is a means to achieve desired end result through process transformation. Client expectations continue to soar high with the advent of new technologies, and many services are embracing process transformation in an effort to remain relevant and fulfill client expectations.

Path breaking buisness enterprises acknowledge that in order to remain competitive, they must constantly improve their culture, processes, data and innovations; attaining this requires extra planning and effort. Several enterprises are still reluctant to earnestly improve their internal process in spite of the established effectiveness of automation and digital transformation. Manufacturing enterprises have a solid understanding about process streamlining besides automating the process to remain competitive. One of the greatest influences of process transformation most likely originated from innovation assisting, not changing or removing workers. These benefits will just be recognized when automation is securely incorporated with organization procedures at all degrees of the organization and also customized to various functions within the company.

For manufacturing enterprises, process transformation is the basis of Industry 4.0, the secret to updating and enhancing manufacturing operations. Transformation impacts new product growth segment in a considerable number of ways. It drastically changes the enterprise's new product landscape. From obtaining basic materials, to bringing products to the market, technology is now becoming omnipresent for producers; modern technology is being incorporated up and down the production line to provide information and determining real-time choices throughout the procedure; thus, manufacturing is undergoing a fundamental change. It offers remarkable possibilities for product developers in the production industry, not just to distinguish their new products but also to improve the method these products are developed, created and released. The paybacks are considerable, but so are the challenges in getting process automation to process transformation right.

CHALLENGES TO INCUR

Automation of business processes is at the heart of process change initiatives and is a crucial parameter for success, although there are still challenges and demands to be taken care of. Most of the procedures are significantly complicated and consist of various actions and elements. Absence of a clear vision besides tactical oversight increases the chances that vital service processes are mishandled, delayed or damaged recklessly generating problems and affecting credibility. Few main challenges are as follows:

- Comprehend the need for process automation within the enterprise, even before preparing for a process transformation. This is essentially imperative as most stakeholders in the organization may not be prepared for a transformation. People have various point of views, and although collaborations succeeds at the outset, it might hold hidden reflexive conflict.
- As consumer demands continues to grow, adapting to the marketplace may require a lot more investments that it might go beyond the actual spending plan set previously.
- A well-laid out strategy requires a vision for the process transformation to improve the existing core expertise and advance toward that vision.
- The transformation to become a customer-oriented enterprise requires constant upgrading of the process.
- Understanding the impact that process transformation can have on an organization.
- Resistance with respect to adopting new technologies that assist in process transformation.
- Defining the metrics that can properly gauge efficiency prior to, during, and after the process transformation.
- A lot of the challenges in attaining process transformation objectives are social and behavioral.

Adopt the strategy of continuous progress in addition to stepwise progress toward process transformation. While one can tackle the challenges as and when they arise, it is also prudent to prepare beforehand to accelerate the ROI and ROV.

VALUE DRIVERS OF PROCESS TRANSFORMATION

The transformation of a process in a manufacturing enterprise includes automating various jobs in order to complete and in order to guarantee its survival. Developments in electronics and technology have actually taken the world of economy and business by storm. Automation will be favored for recurrent manual tasks and the costs of robots besides their control systems will decline also in the future so small and medium sized manufacturers can take advantage of them. Product development organization will start making use of subtractive and additive processes and will certainly have a deep understanding of naturally derived frameworks and digital tools. Production operation will be extra carefully tied to either the place of the resources or the location of the consumer. Wearables and powered exoskeleton will boost human performance and improve safety. Smart machines are progressively functioning effectively in the process sector and are expected to perform many routine jobs in production and warehousing. Manufacturers are having a hard time drawing in the ideal sort of ability they need as it has in fact become very sophisticated and extremely progressive in numerous ways.

Smart manufacturing ability to check and monitor online makes it possible to bring data-driven choices to human tasks to enhance the performance of systems and processes and save time. Eliminating shop floor operation concerns during the development process by making use of Industry 4.0 bars such as advanced statistical process control (SPC) and electronic performance administration can be beneficial. Quality and speed are the first things that are enhanced in this connected environment, which, in turn, assures something that all suppliers desire: an improved client service experience. Copyright protection is another main challenge in an increasingly distributed worldwide manufacturing ecosystem. Process standardization lowers the price of entry for suppliers to carry out even more kick start process improvement, and these drivers interrupt and transform the process in the production sector irrevocably.

SUMMARY

Organizations across the globe are changing quickly, driven by emerging digital innovations. Manufacturers of small, medium and large enterprises use process automation together with process transformation, which is likely to become the standard. Transformation in the manufacturing sector is commonly comprehended to mean the adoption of electronic technology to replace or automate manual procedures. Product manufacturers, today, deal with a transforming business standard, in which emerging modern technologies are permanently transforming how products are manufactured and service is provided. With unprecedented data availability, product options and network alternatives, consumers are demanding an ever-increasing level of transformation, not simply in products and services but also across the whole procurement and product use experience. To get rid of delays, minimize accidents, eliminate mistakes, improve product quality and develop new organization standards, automation innovations are increasingly imperative in today's manufacturing industry. The race to determine and also cater to ever-changing client requirements is getting intense with the advent of new players with unique business models. Process transformation indicates transforming standard procedures in to more efficient digital systems that can boost performance dramatically, improving all aspects of the procedures.

Thanks to transformation of manufacturing industries, the manufacturing facilities of the future will be more effective in the utilization of robots, material and renewable energy along with human resources. Services and product improvements suggest developing new value-added services that can both boost the production environment and the consumer experience while opening brand-new revenue streams. Major changes on the demand side are also happening with increasing transparency, consumer involvement and brand-new patterns of customer habits, which are increasingly built upon their accessibility to mobile networks and information, pressure manufacturing enterprises to transform the method they create, market and provide the products along with services. New modern technologies make products much more resilient, while information besides analytics is changing exactly how they are maintained. Manufacturers will begin exploring the journey in the midst of the inexorable transition from easy digitization to technology based upon combinations of innovations through the collaboration of human knowledge with bots toward a future that reflects typical objectives as well as values compelling the firms to review the means of how they work.

BIBLIOGRAPHY

Aras| Product Brief | Product Lifecycle Management, https://www.aras.com/en/capabilities/product-lifecycle-management.

Bizagi. N.d.Digital Process Automation. https://www.bizagi.com/en/solutions/digital-process-automation

DPA | Product Brief | Opentext, https://www.opentext.com/products-and-solutions/products/digital-process-automation.

Elangovan, U. 2020. *Product Lifecycle Management (PLM): A Digital Journey Using Industrial Internet of Things (IIoT)* (1st ed.). New York, CRC Press. Doi:10.1201/9781003001706.

Haigh, M. J. 1985. *An Introduction to Computer-Aided Design and Manufacture*. Oxford, UK, Blackwell Scientific Publications, Ltd., GBR.

Hofstede, A., W. van der Aalst, M. Adams, and N. Russell. 2009. *Modern Business Process Automation: YAWL and its Support Environment* (1st ed.). Berlin, Germany, Springer Publishing Company, Incorporated.

IPA | Product Brief | Uipath, https://www.uipath.com/rpa/intelligent-process-automation.

Leon, A. 2014. *Enterprise Resource Planning*. New Delhi: McGraw-Hill Education (India) Pvt Ltd.

Luther, David. 2020. 21 Ways to Automate a Small Business. https://www.netsuite.com/portal/resource/articles/accounting/small-business-automation.shtml

PLM | Product Brief | https://www.3ds.com/products-services/enovia/products/.

PLM | Product Brief | https://www.autodesk.com/content/product-lifecycle-management.

PLM | Product Brief | PTC, https://www.ptc.com/en/products/plm.

PLM | Product Brief | Siemens PLM, https://www.plm.automation.siemens.com/global/en/products/plm-components/.

RPA | Product Brief | Automation Anywhere, https://www.automationanywhere.com/rpa/robotic-process-automation.

RPA | Product Brief | UiPath, https://www.uipath.com/rpa/robotic-process-automation.

Scholten, B. 2009. *MES Guide for Executives*. Research Triangle Park, NC: International Society of Automation.

Watts, Stephen. 2020. The Importance of Digital Process Automation (DPA). https://www.bmc.com/

Zeid, I. 1991. CAD/Cam *Theory and Practice* (1st ed.). New York, McGraw-Hill Higher Education.

3 Technological Innovations of Industrial Revolution 3.0 to 5.0

Manufacturing technology pledges to influence every facet of the production services from design, research and development, manufacturing, supply chain to sales and marketing. Industrial manufacturing sector processes raw materials into new products, which are eventually utilized by customers, and also, it is influenced by the commercial transformations that are taking place rapidly. Technology along with innovation is the vital driver of advancement in manufaturing besides performance improvement. New technologies combined with cutting-edge product designs provide manufacturers numerous chances to improve the core business value, especially for small and medium sized enterprises. To improve product development, manufacturers can engage different modern technology tools as the first step toward digital transformation.

INDUSTRIAL REVOLUTION

In the contemporary period, manufacturing facilities are often transformed by technology-based industry hubs. Trade expansions were made as a result of the transformation from an agricultural economy to an industrial machine-driven economy beacuse the automation of design and manufacturing of products through services resulted from the quick development of the technology across different industrial sectors. The primary revolution that occurred during the industrial transformation was the renovation, execution and adoption of technological innovation. The recent development of technology innovation is a recurring journey; the speed of innovation and transformation continues to enhance. Transformation is not new to industrial sectors; it is considered as a significant resource of industrial economy. Organizations run by business leaders' with a clear vision toward digital transformation upgrade their business models to promote business growth, stay relevant in changing times, and distinguish themselves from the competition, being able to think artistically and welcome innovation right into their product lines.

Let us have a look at the brief overview of different industrial revolutions.

FIRST INDUSTRIAL REVOLUTION

Changes commenced at the end of the eighteenth century and continued until the beginning of the nineteenth century in the industrial sectors in the form of automation. The first industrial revolution was a major turning point in the background of the human culture and was also called as the age of mechanical manufacturing,

DOI: 10.1201/9781003190677-3

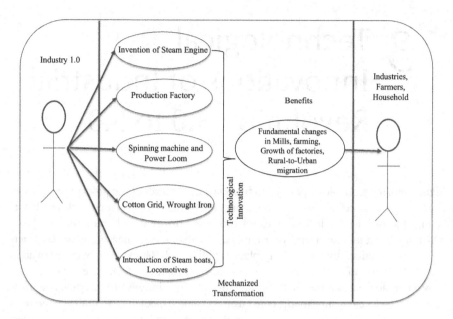

FIGURE 3.1   Mechanical revolution.

where products made from hand tools were transformed to products that were made by machines. With the invention of the spinning jenny and the use of power began the era of industrial revolution. Garments were made much quicker than ever before. With the advent of the steam engine, steam power powered everything from agriculture to apparel manufacturing. The product business, especially, was changed by automation, as was transportation. Power generated from heavy steam and coal made manufacturers discover new industrial use. The steam-powered locomotive revolutionized the transport of goods. Additionally, it assisted in the development of cities, and the quick expansion of, consequently, the economy expanded together with them.

> James Watt's advancement consisting of a separate condenser significantly enhanced heavy steam engine efficiency besides incorporating a crankshaft and gears became the model for all modern steam engines. His discovery was considered one of the most effective creations of the Industrial Change.
>
> *(Famous Scientists, n.d.)*

It was more affordable to have a machine than having individuals that would have to be waged. There occurred a paradigm shift from an agriculture-based economy toward machine-based production. As a result of these advantages, for the right or for the worse, man-made machines are all still being utilized as we speak. Throughout the industrial transformation, environmental pollution increased because of the use of the new fuel, the development of large manufacturing facilities and the surge of unsanitary metropolitan facilities. The first industrial transformation was a time that initiated lots of socioeconomic reforms together with several of the most functional technological wonders.

SECOND INDUSTRIAL REVOLUTION

Numerous technological innovations in the industrial sector aided the introduction of the internal combustion engine, innovation of electrical energy, use of steel, chemical sectors, alloys, petroleum and electric interaction technologies. The current manufacturing and the production techniques of the first industrial revolution were improved by the technological transformation, which marks a stage of rapid automation. Advancements in production technology, materials and manufacturing tools are made to standardize all kinds of goods made in the manufacturing landscape. Science and technology innovation work together, where science reveals effective insights that drive technical development. Numerous innovators, understanding the benefits of internal combustion over heavy steam, tried their hand in utilizing energy in the automobile field. The development of continuous flow processes with interchangeable parts, usually related to production line, led to the automobile industry setting up a plant to standardize mass production of intricate products.

> Invention of electric light by Thomas Edison made a huge impact in the manufacturing sector, made the operations to run three shifts per day. *Technology of three-axis system by Wright brothers, that made the airplane to maintain security also steadiness, happens the basic principle continues to be the identical likewise today in the aviation field. The electric generator technology adds to modern family products such as refrigerators and laundry equipment's, in addition with the innovation of internal combustion engine enabled both automotive as well aviation field.*
>
> *(Stanley, 1931)*

Telephone, radio, conveyor belts, cranes and machines are all powered by electrical power. Likewise, hydroelectric power plant and coal-based steam power plants were

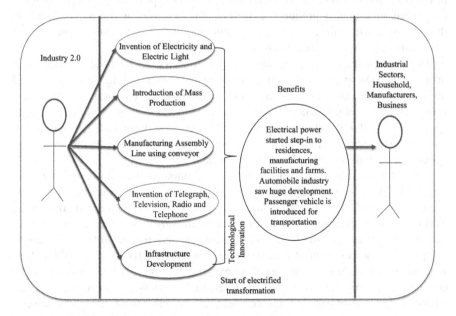

FIGURE 3.2   Science and technology revolution.

developed. Steel use increased in the place of iron, in the construction of ships, high-rise buildings and larger bridges. The influence of the industrial revolution on financial development and performance was far more absolute than any type of technical advancement and it contributed to the global merger of markets coupled with the first industrial transformation. Thanks to all of the developments and inventions, the second industrial transformation can be summarized as a positive and advantageous time in history. The inventions of the electrical energy, automobiles, and aircraft in the beginning of the twentieth century are some of the reason for the second industrial transformation to be considered as one of the most vital ones.

Industry 2.0 introduced processes leading to improved product quality and manufacturing effectiveness and efficiency such as just in time (JIT) and lean principles to enhance Industry 1.0. Equally, first and second industrial transformation made conspicuous contributions for the development of the industrial sector. We cannot reject on the fact that automation and industrial transformation have caused some unfavorable effects to the globe. Nevertheless, endorse the words of Heraclitus – change is the only constant in this world.

THIRD INDUSTRIAL REVOLUTION

The second half of the twentieth century saw a surge in computer systems, automation, robotics, renewable energy, nuclear energy, electronic devices, telecoms, the Internet and digital revolution, what is called as the third industrial revolution. In the second half of the twentieth century, the industries were often affected by technological advancements and unpredictable market variations and international competitors. Typically, a manufacturing industry in order to sustain itself and be relevant in the competitive world of production needs to make production systems that not only create high-quality products at cost-effective expense but also adapt quickly to the marketplace transformations and consumer requirements; besides, the production environment should have absolutely no machine downtime.

The implementation of automation during Industry 2.0 and the automation of manufacturing shop floor lines with the digital uprising was a significant leap onward in Industry 3.0. Industrial machine manufacturers started improving the performance of the machines incorporating integrated circuits. The era of automation was spearheaded within the automotive sector it was eventually embraced throughout all manufacturing sectors. Industry 3.0 promoted the growth of software application systems to capitalize on the electronic hardware.

Electronic devices along with information technology started to automate manufacturing and also take supply chains global. Complex and recurrent jobs were performed by software programs, making it possible for process automation in the workplace, in the product design and in the manufacturing. Innovations in computer assistance, the advancement of microprocessors and the advantages related to computerized process control were recognized in the industrial sectors. While there were many substantial technological innovations throughout this period, it is the emergence of computer-assisted applications, such as computer-aided design, computer-aided manufacturing, computer-integrated manufacturing, computer numeric control, enterprise resource planning, material requirement planning, customer relationship management, supply chain management, rapid prototyping, product lifecycle

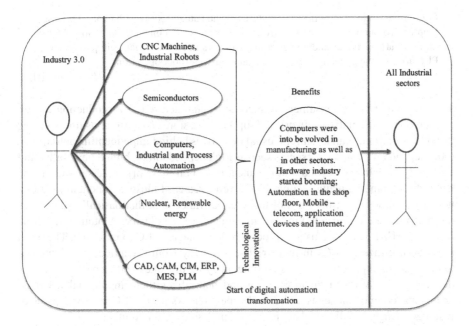

FIGURE 3.3   Digital automation revolution.

management, manufacturing execution systems, programmable logic control, supervisory control and data acquisition (SCADA), which enabled the NPD/NPI teams to strategize routines and track product streams through design and development within the manufacturing process.

Programmable Logic Controller

A programmable logic controller (PLC) is a computer particularly made to run under severe industrial settings such as extreme temperature levels, damp dry, and dirty conditions. It is a compact industrial computer system created to manage manufacturing system process from one place. Simply put, it is an industrial solid-state computer that keeps an eye on inputs and outcomes and makes logic-based decisions for automated processes as well the shop floor machines in the assembly line. PLCs play a critical role in the field of automation, making use of creating part of a bigger SCADA system. The range of PLC features are timing, counting, determining, comparing, and handling different analog signals. PLCs are usually referred to as industrial computers.

Technical breakthroughs in handling communications are generating new avenues for industrial automation. Regardless of the fast developments in modern technology, PLCs remain to play an important function in production and also work as a central processing unit for all actual time decisions. PLCs help in identifying, monitoring and getting rid of waste and also come to be more recognized as the world of automation control caters to the world of modern-day information and communications technology. Supervisory control systems play a vital role in the manufacturing industry as a move towards more open design standards, which enable interaction between conventional PLC systems and other manufacturing control systems.

PLCs are used to control detailed features of the batch process with continual closed-loop control along with extra controllers supervises the complete operation. As well used to manage linear and also rotating actuators in an industrial fluid power circuit. PLC are suitable gadgets for regulating an industrial robot operation.

(Laurean, 2010)

The vast majority of consumer products are manufactured at a production center, delivered via a circulation network and supplied to a store and to the consumer making use of automation. Manufacturers' usual objectives behind implementing automation tools depend on a couple of aspects such as high dependability, high repeatability and simplicity of delivery beyond maintenance. Based on these principles as well as the need of the manufacturing section, PLCs were created. Industrial automation in this new age is practical in the form of technical advancement that basically supervises the performance of various kinds of machines in the industry. A human–machine interface (HMI) enables a user to productively manage a PLC in real time. The main benefits of utilizing PLCs in industrial sector array from exceptional performance to decrease in expenses. PLCs assist in identifying mistakes and also detect quality deficiencies early on in manufacturing and neutralizes them intentionally. Cross-checklists, component scans and completeness checks using PLC are some steps of manufacturing surveillance that ensure optimal quality in manufacturing.

SCADA

SCADA is an automation centralized control system that checks and regulates whole sites, ranging from an industrial plant to a complex manufacturing plant. Industrial organizations began to use relays and timers to maintain some degree of supervisory control without needing to send individuals to remote places to interact with each tool. Increased usage of microprocessors and PLCs increased the business' capability to monitor and manage automated processes. With the adoption of modern ICT requirements, today's SCADA enables real-time plant details to be accessed from anywhere across the buisness enterprise.

SCADA allows an organization to meticulously research and anticipates the optimum action to gauged conditions and implement thoses actions immediately each time. SCADA systems make use of distribution control systems (DCS), process control systems (PCS), PLC and remote terminal units (RTU).SCADA assists in reducing manufacturing waste and boosting the overall performance by providing relevant and comprehensive production information to the drivers and also the administration. SCADA handles parts lists and for just-in-time manufacturing as well as controls industrial automation and robotics. It ensures high quality in addition to process control in production shop floor.

A common inquiry might arise in the minds of small and medium manufacturing enterprises, who are new to automation is whether both PLC and SCADA are the same or different and how they needs to be utilized within the enterprise. Two of the most important technical advancements within manufacturing industries are SCADA and PLC. Both the technologies work together to offer essential services. PLC is a physical equipment, whereas SCADA is a software application. SCADA is made to operate in a much more comprehensive range given that it can check and gather information from every result of a system, whereas PLC, on the other hand, focuses on keeping an eye on just one aspect within the system.

SCADA helps companies enhance their operational effectiveness. It refines real-time information so that the control team has up-to-date, precise information to make smart decisions. It manages operations, enhances effectiveness and reduces downtime. Moreover, the system provides sophisticated warnings and effective maintenance, helping manufacturers to minimize damage.

Industrial Robots

Automation has been adopted by many industries. The advancement of machinery on top of other technological developments replaced manual labor. The advancement of numerically controlled (NC) devices and the increasing popularity of the computer both led to the generation of the first industrial robots. Robots have become an essential part of today's huge manufacturing industries; they are microprocessor-controlled and smarter besides having a greater degree of operational flexibility. Competitors from firms faced a high need for commercial robots. So, what is a robot? I layman's terms, a robot is a machine that is capable of carrying out regular and also intricate actions that are set by engineers.

Industrial robotics has the ability to considerably improve the quality of the product and also form an inalienable part of the contemporary production landscape. Applications are performed with accuracy and efficiency. As machines continue to develop and handle increasingly complicated tasks, all manufacturing procedures will soon be automated and taken over by robots. Most of the enterprises – SMEs or OEMs – are incorporating the use of robotics in their contemporary manufacturing facilities.

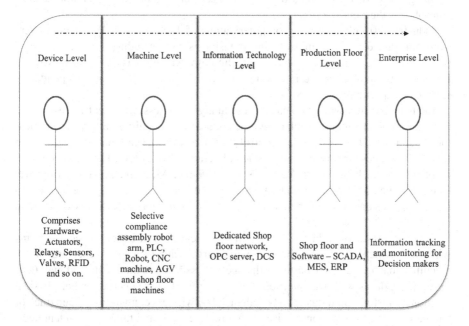

FIGURE 3.4 Automation levels.

The most frequently utilized industrial robots are selective compliance assembly robot arm (SCARA), articulated, Cartesian, gantry and delta robots. Robots are the future of production, and they provide suppliers increasing opportunities to minimize prices, boost manufacturing and remain affordable.

The first generation of industrial robots had limited intelligence, freedom and also functional levels of liberty. Human beings might experience exhaustion as a result of the recurrent nature of their work, which can cause them to make errors. Robots, on the other hand, can stay clear of making such blunders as a result of their dexterity and high degrees of machine learning. The impact of automation in production spreads far and wide, boosting efficiency and success of the whole manufacturing enterprise. Robots positively influence production by taking on recurring jobs, streamlining the general setting of operations and teaming up with humans for product manufacturing. Even small to medium sized enterprises are realizing the significance of integrating robotics into their process for long-term benefits.

Fourth Industrial Revolution

The application of computers and automation in Industry 3.0 opened new possibilities for advancement in industries with smart as well as self-governing systems fueled by data and machine learning, what is being called now as Industry 4.0 or the fourth industrial revolution. Today, in the smart connected world, automation no more indicates stand-alone robots running independently of each other rather; industrial markets are seeing much more robust, automation solutions, which take advantage of big data, the industrial Internet of things (IIoT) and data analytics. By integrating software and hardware, manufacturers can maintain a comprehensive control over their entire operation.

Industry 4.0 enhances the computerization of Industry 3.0. When computers were introduced in the manufacturing industry as part of Industry 3.0, the process of incorporating a brand-new modern technology was not easy for many. Now, in Industry 4.0, computers are interconnected with each and function independently sans human participation.

Industry 3.0 find ways to measure and analyze processes to identify restorative actions for enhancement, which paved the course to utilize highly efficient statistical tools, such as Statistical Process Control (SPC), and the seven Quality management tools. The process involved information collection, which permitted the techniques of plan–do–check–act (PDCA), six Sigma (Define, Measure, Analyze, Improve, Control) to be used, which can assist manufacturing industries to improve their existing techniques and advancements and also make them more effective with the help of Industry 4.0 technologies.

The real power of Industry 4.0 lies in the integration of cyber-physical systems with IIoT, which makes the smart manufacturing facility a fact. Emerging smart machines that are getting smarter as they get accessibility to more data and manufacturing procedures will become even more effective. To dive deep in, cyber-physical solutions transform an industry to be enhanced by wireless connectivity, keep track of and monitor the business enterprise from a remote place and take independent decision, thereby adding a new dimension to the production process. Machines, people, processes and frameworks are integrated into a networked loop enabling the overall monitoring to become highly reliable and a lot more streamlined.

Industry 4.0 is creating a new business value by increasing the outcome, asset use, besides overall efficiency. It is not merely acquiring new technology and systems to enhance the manufacturing performance: it is transforming the process by which the manufacturing industry operates and expanding their presence across the globe. The outcome of Industry 4.0 is that the cross-functional team (CFT) of the organization shares refined, up-to-date, pertinent views of manufacturing along with business process with lot more dynamic analytics.

Industry 4.0 is ready to take root throughout the manufacturing environment. By understanding and also utilizing the modern technologies, manufacturing can industries journey toward digital transformation. It is the merely height of technological innovation in manufacturing, but it still sounds as if machines are taking control of the industry. It is absolutely a cutting-edge method of manufacturing technology, which ensures manufacturers a new degree of optimization and efficiency.

Automation in the industrial sectors has progressed from making use of basic hydraulic and also pneumatic systems to contemporary robotics. IoT/IIoT can be applied in the manufacturing systems to make the shop floor operation simple with an automated control system. Process automation advances from reducing human interference in the systems to preventing carcinogen and enhancing efficiency and effectiveness. Having successfully incorporated the technical advancements in the past, the wireless-based automation system is being embraced in the various types of systems in addition to making use of Industry 4.0.

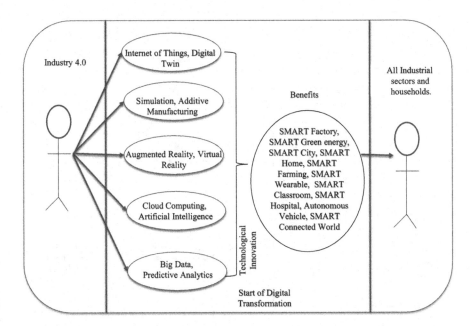

FIGURE 3.5   Cyber-physical revolution.

For manufacturers across the globe, Industry 4.0 stands for a paradigm shift in just how are industries run, as essential as the transformation from Industry 1.0 via Industry 3.0. Without manufacturing, the economy of any country will absolutely stumble, and it depends on the manufacturers to provide improved capacities and optimal techniques to produce products. Leading key digital technologies associated with Industry 4.0 are explicated as follows.

Internet of Things

Extending the power of the Internet beyond computer systems and also smart devices to an entire range of things, processes, and environments made a huge impact to both the industrial and business sector. When a physical thing is linked to the Web, it implies that it can send out details or receive details, or both. The capability to send out and/or obtain information makes things wiser and also smarter. As a whole, Internet of things (IoT) is a network of uniquely recognizable things that interact without human interaction, generally making use of IP connectivity. The semantic origin of the expression is composed by a couple of words: "Internet" as well as "Thing," where "Internet" can be specified as "The worldwide network of interconnected local area network, based on a conventional interaction process, the Internet collection (Internet protocol suite – Transmission Control Protocol/Internet Protocol TCP/IP)," while "Thing" is "an object not exactly recognizable".

> One of the best examples of IoT is remotely manage the on/off the lights as well monitor water level in the overhead tank by means of your SMART mobile phone without being physically present.

(Elangovan, 2019)

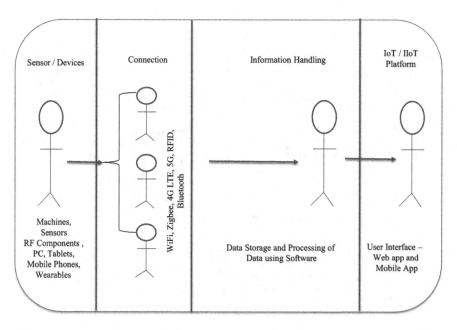

FIGURE 3.6 Function of IoT.

In a nutshell, IOT is the inter-networking of physical devices, connected devices and also smart gadgets, buildings, and various other products installed with electronics, software application, sensing units, actuators, and also network connection which enable to collect as well as exchange data. The idea of including sensors and intelligence to fundamental objects was reviewed in the 1980s through 1990s. Few developments were introduced as Internet-connected vending machine was listless since the innovation was not ready. Processors that were cheap and power-frugal enough to be just about non-reusable were needed before it ultimately became economical to link up billions of tools. The implementation of radio-frequency identification (RFID) tags low-power chips, which can connect wirelessly, solved several issues, in addition to the enhancing the availability of broadband along with cordless networking. Adoption of IPv6 that, among other points, must offer enough IP addresses for each device across the globe is also a necessary step for the IoT.

Industrial Internet of Things

Industrial Internet of things (IIoT) has become prevalent in the industry as digitization has become the manufacturing enterprise's top priority. Thus, IIoT is a subcategory of the IoT, which consists of consumer-facing applications such as wearable devices, smart factory, autonomous vehicles and robots. Sensors embedded in the shop floor machines transmit data by means of intranet and are run software programs that are the characteristic of industrial IoT. IIoT is changing the means industrial companies operate daily by incorporating machine-to-machine interaction with large information analytics in real time, and the IIoT can aid companies recognize their organization procedures much better by assessing the information originating from sensors, making their processes extra effective and open to new income streams. An IIoT ecosystem is where individuals, applications and devices connect. That is why most large industrial IoT services are based around a main industrial IoT system that can manage every facet of the commercial IoT network and also the data streaming via it.

The key distinction between IoT and IIoT depends on the business application. Industrial IoT centers around real-time collection besides evaluation of granular information from linked sensors, allowing fast renovations to operational effectiveness and also efficiency, instant stock control over and significant cost savings.

Basic components that are required for IIoT are as follows:

- Hardware
- Software
- Processing unit
- Cloud
- IIoT platform – System App, Mobile App

Manufacturing enterprises are eager to carry out connected factory, Industry 4.0 and IIoT concepts to realize benefits, such as minimized operational prices, better exposure, control, and operational efficiencies. These benefits can be accomplished by a variety of means, one of which is making use of data gathered from keeping an eye on factors along a production line to reduce waste and downtime. As a crucial element of digital transformation, adopting IoT technology is imperative to the manufacturing

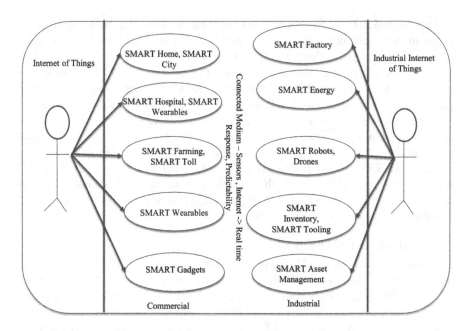

FIGURE 3.7   IoT vs IIoT.

sector. As the international market pushes manufacturers to reassess operations, smart manufacturing powered by IIoT-driven data analytics is necessary. Digital transformation of a conventional production procedure into a smart manufacturing facility will be certainly advantageous for manufacturing enterprises of all sizes.

3D Printing

Additive manufacturing (AM) is a 3D printing process that constructs 3D products by adding layer upon layer of the product according to digital 3D CAD model information. AM was originally used for quick prototyping, namely to make visual and useful prototypes. It can considerably quicken the product development process. 3D printing constructs a model in a container full of powder of either starch- or plaster-based product. An ink-jet printer head shuttle uses a percentage of the binder to create a layer. Upon application of the binder, a new layer of powder is brushed up over the previous layer with the application of even more binder. The procedure is repeated till the model is complete. As the model is sustained by loosened powder, there is no requirement for support. Furthermore, this is the only process that builds in colors.

AM opens brand-new opportunities in challenging markets such as the healthcare, automotive, aerospace sectors, consumer goods together with commercial production. 3dprinting assistances on-demand manufacturing organization model stresses the price of delivery as well the ability to produce extra components much faster in enhanced manufacturing uptime and also much less production disturbances at the point of need.

(Gonzalez, 2021)

Design for manufacturing (DFM) frequently shows that designers need to customize their designs to fit restrictions related to the conventional production treatments in order to make sure the expediency of building the model. Nonetheless, this might lead to restrictions and constraints in the designers' innovative flexibility for new product development. Conventional manufacturing techniques can produce a wonderful range of designs; nonetheless, 3D printing takes manufacturing to the next level. Among the greatest advantages of this modern contemporary technology is the greater series of forms and shapes, which can be developed.

Augmented Reality

Augmented reality (AR) improves the physical world around us with the help of modern technology. Innovation superimposes information along with online things on real-world scenarios in real time. It makes use of the preexisting environment and includes information to it to make a new artificial environment. It superimposes digital information and photographs on the real world, promises to shut the space and launch untapped, along with distinctively human abilities. AR applications are provided with mobile phones, but progressively distribution will move to handsfree wearables such as head-mounted displays as well smart glasses.

> Utilizing AR technology for remote upkeep would permit any type of employee with an AR device to be guided on a machine malfunction on the production floor by a professional located at his premises. Microsoft's HoloLens mixed reality headset, a mix of AR and virtual reality modern technology, has already been used by few manufacturers.
>
> *(Microsoft, n.d.)*

Safety is constantly a concern in the manufacturing environment. In manufacturing, the modern technology can be used to determine a range of changes, recognize risky working problems or even envision a completed product. Wearable AR tools for manufacturing shop floor workers lay over production setting up and also solution directions, and it is supplementing traditional guidebooks, cookbooks and training methods at an ever-faster rate. AR will definitely reinvent manufacturing, and businesses need to offer it major factors to consider.

Data Analytics

Data analytics in manufacturing is focused on collecting and evaluating information instead of process control. Data from an endless variety of sources such as ERP, MES and machines can be gathered as well as associated together to recognize areas for enhancement. Improving the top-quality product by minimizing process variation has always depended on data. To lower functional risks as well as improve service performance by leveraging advanced data analytics such as statistical analytics, predictive analytics, and so on are very crucial questions that need to be considered, currently in the smart connected manufacturing.

So, what is data analytics? It is the method of collecting insight by breaking down past efficiency and also information to make sure that an insightful next step can be intended and also taken. It describes the collection of measurable and qualitative techniques for obtaining beneficial insights from data. One of the straightforward examples I can think of is usage of data analytics by Amazon.com. It utilizes data analytics to recommend the best item to the customer based on the item that they bought in the past.

SPC analysis uses the ability to boost the product quality and improve the process efficiency and effectiveness, which is something that every manufacturing organization demands. Importance of data analytics in manufacturing operations cannot be overemphasized. SPC is the keystone of quality assurance in manufacturing process. Over the years, suppliers have used statistical devices to research historical data to expose details relating to special differences between equivalent things: shifts, items, devices, procedures, plants, great deal codes and more. When evaluating processes, it is very vital to compare common causes along with the unique root causes of the variant. Special sources of variation show a process modification, which requires to be examined.

The need to accurately forecast demand is critical to the manufacturers. Analyzing demand in real time is inefficient given that companies need to make decisions about the demand ahead of time to complete a whole production cycle and deliver the end product to the customers. With predictive analytics, it is feasible to not just boost the manufacturing quality, increase return on investment tool and overall equipment effectiveness (OEE) but also prepare for various needs throughout the business, exceed the competition, and guarantee consumer safety.

Production ventures have process professionals, operational excellence teams, as well as designers who are smart and capable with an intimate understanding of the production procedure, yet they need easy and instinctive logical devices to pull the value out of information. Path to supplying impactful data-driven production jobs is loaded with possible obstructions and mistakes. By encouraging process designers with advanced analytics tools, more production issues can be assessed by analyzing the information. The fostering of big data, machine learning, robotics, artificial intelligence (AI) and IIoT is greatly impacting the industry and company.

Simulation

Simulation has become a part of NPD across industrial sectors allowing the product or component or total system habits to be discovered and also tested in a virtual environment. Simulation has actually established a close relation with both the computer system industry and product design processes; it also provides an inexpensive, protected and fast evaluation tool. Finite element analysis (FEA) is the simulation of a physical sensation utilizing a numerical and mathematical technique referred to as the finite element method (FEM); advancement in computing, modern programming language, visualization tools and graphics have actually had a significant influence on the development of simulation innovation. Real-time simulation technology is made use of today in different industrial sector applications such as manufacturing, energy, power systems, industrial products, valves, pumps, automotive and aerospace. It is well-established during product design and validation that using simulation methods to manufacture shop floor such as installing new manufacturing centers, assembly line and also procedures yield huge advantages.

Simulation assists product design group, and a large range of various digital versions of the item can be developed and examined, making simulation part of the design process itself. Another benefit of simulation is the possibility of carrying out screening remotely from any part of the globe, which has proved to be a blessing during the COVID-19 pandemic. Simulation has become essential enabling the technology of Industry 4.0 in decision-making, design as well procedure, covering the entire life cycle of a production system and also paving way for the development and implementation of Industry 5.0 to increase the effectiveness, safety, security and ecological demands. Simulation is the only way to attain the intricacy of modern-day product design in control and to successfully make use of the possibilities offered by a quickly implemented technology. On collaborating with AM, simulation makes sure that the final component not only has the optimal form but can also be produced specifically, cost-effectively with a high degree of consistency. Simulations on the digital twin can provide other crucial product information such as the do's and don'ts for optimal efficiency, forecast important failings and maintain requirements.

FIFTH INDUSTRIAL REVOLUTION

The future of the industries is all about progress. Industry 4.0 is still the most preferred technology among the manufacturers. The manufacturers of small and medium sized enterprises have either partially or completely incorporated Industry 3.0 and Industry 4.0, and are eager to introduce more technological improvements. Hot progression in AI, robotics, ML, data analytics, among others, is causing the birth of the fifth industrial revolution or Industry 5.0. It will be an AI transformation with other innovations such as quantum computing and integrating people, process, machines and environment with each other. Robotics is becoming more vital as it can now be paired with the human mind using advancement in AI. A strong need to increase the productivity while not removing human employees from the manufacturing industry is imposing new challenges on the global industrial economic situation. Industry 5.0 will integrate humans and machines to exploit the human mental ability and creativity even better and to improve the process performance by integrating process with smart systems.

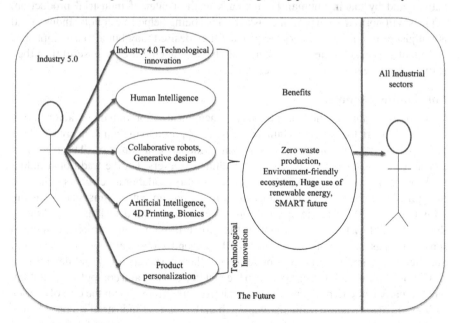

FIGURE 3.8 The future of the industrial economy.

Industry 5.0 integrates intelligent automation, gadgets and systems at the work environment to increase cooperation along with collaboration between people, process, robots and shop floor machines. It assists highly skilled employees to lead smart devices and robots to work far better. Economy and environment might see significant influences because of reduced waste product as manufacturing enterprises target zero-waste production, lowering material and waste management costs. In regard to the social environment, Industry 5.0 will certainly lay greater emphasis on the human aspect of manufacturing, whereas Industry 4.0 concentrated only on the technology innovation.

> One real-world instance is used by FANUC, a Japanese robotics business that's a pioneer in lights-out production or dark factories they're geared up with completely automated systems that can function in the dark without human guidance.
>
> *(Wheeler, 2015)*

Connecting the virtual and physical worlds is the main criterion for the manufacturers to examine data, keep track of the manufacturing process, handle risks and reduce downtime; all achieved by simulations with the advent of digital twins. With the current innovations in large data handling and AI system, it is currently possible to create a lot more sensible models depicting various operating circumstances and also characteristics of a process. While representing unpredictability in the process, digital twins offer an immense possibility by enabling reduced wastefulness by collaborating with the system. Industry 5.0 will bring unmatched challenges in the field of human-machine interaction as it will certainly place machines extremely close to the day-to-day life of any human.

Industry 5.0 uses the innovation established in Industry 4.0. Enabled by innovations and by placing human beings back at the center of industrial production, devices will normally perform the tasks besides being helped by cobots. Industry 5.0 is not just providing consumers the product they desire today but also accomplishes tasks that skyrocket to new elevations and also are much more purposeful than they actually have been in more than a century.

Collaborative Robots

Collaborative robots, called as cobots, are a new incarnation of a manufacturing bot designed to work together with human beings as opposed to in their own area. Cobots are experiencing rapid market development in industrial automation. These are created to function flawlessly, together with human workers. Unlike traditional industrial robots that might hurt a person in their vicinity, collaborative robots make use of sophisticated aesthetic technology and are geared up with sophisticated sensing units that allow them to identify individuals and change their task as well. Among the greatest safety function of cobots is their force-limited joints, which are made to sense forces as a result of impact and swiftly respond. Cobots are beneficial to small and medium manufacturing enterprises due to their cost, versatility and flexibility.

Cobots are gaining popularity due to the fact that sensors and computer technology have actually come to be so inexpensive that they are driving down the cost of robots. Cobots are also easier to train and deploy than the massive industrial robots. Cobots can be utilized is in every industrial manufacturing process from fabrication and product packaging to CNC machining, molding, testing, quality assurance and so on.

Cobots will not replace human employees, rather they are going to work along with them, accomplishing repeated jobs, which will certainly free workers to pursue other tasks. Let us have a quick look on the evolution of cobots.

Robots were considered as machines of the future in the early 1980s; manufacturers began pressing the frontier onward to sustain industrial growth and achieve greater production competitiveness by incorporating advanced sensing units and primary machine vision systems. Like all advanced modern technologies, cobots were initially met with substantial hesitation by the production industry; one such difficulty was the requirement for fine dexterity and safety. In the early 2000s, growth in industrial robotics was greatly driven by innovations in software application in addition to emerging fields, such as ML and AI. This advanced the frontier of what robots can do giving them the capacity to find out, boost, make decisions quickly to prevent challenges without quitting the production operating at full speed and without any assistance from humans. The initial cobot that can safely operate along with staff members, getting rid of the need for safety caging or secure fencing, was introduced by Universal Robots in 2008. Cobots developed for various applications still call for special safety and security requirements as described by ISO safety requirements and qualifications (ISO 10218). With considerably reduced costs, cobots were a lucrative automation option for small and medium sized manufacturers. In addition, cobots broke all the norms for industrial robotics and consequently amassed widespread attention in the manufacturing industry.

Manufacturers are in actual need of flexible options, cobot-based quality assurance and evaluation systems that can transition between different final products in very little time and end up being very attractive particularly to manufacturers aiming

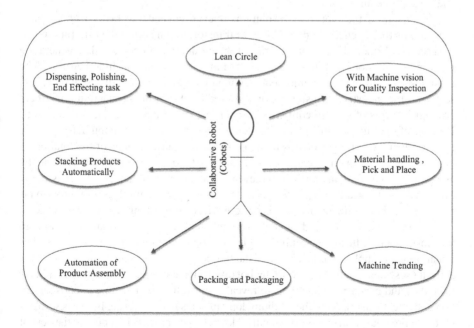

FIGURE 3.9 Applications of cobots.

to satisfy the quality control demands of high-mix, low-volume manufacturing operations. It is necessary for the manufacturer to do some research first, to comprehend their business need, to access the investing capability in as well as understand the innovation they are looking for. Cobots uses smart innovation, with downloadable applications that make it very easy for someone with little or no experience to create a collection of commands for their bots with just a few taps from their tablet, computer or smart device. With the market's value continuing to rise, cobots might soon come a staple in every industrial sector manufacturing ecosystem.

Artificial Intelligence

AI provides the machine the ability to execute a task and reduce human effort with the help of tools and also techniques that were created to provide the machine with the potential to achieve tasks without human interference. AI is a modern technology that can solve a great deal of business or personal activities that need decision-making, intricate reasoning and knowledge. AI is a lasting technological development of the future industrial economy. AI is the simulation of all-natural knowledge in machines that are configured to discover and imitate the actions of humans. AI systems today primarily consist of neural networks that are educated with the help of machine learning as well as deep learning. Virtually, AI systems need to first get the necessary understanding to function. It does not matter whether it is images, texts, language or any kind of data. It is vital that the training data record be refined digitally. AI has acquired thrust; prominent application providers have actually succeeded in developing conventional software applications to create much more alternative platforms as well as options that far better automate business intelligence and analytics procedures.

The advent of the industrial revolutions opened engendered many technological innovations that opened the path to digital transformation across the different industrial sectors. Hundreds of variables impact the production process, as data generating from the shop floor machines are a perfect input for AI and machine learning. The most up-to-the-minute term among the technology tycoons across different industrial sectors is the industrial transformation powered by AI. In the production area, there are digital twins of certain equipment properties, entire machines and components. Because of the shift toward personalization in consumer demand, manufacturers need to take advantage of digital twins to create numerous permutations of the final product. AI aids the maintenance groups to identify potential downtime and accidents by assessing the sensor data attached to the shop floor machines. Industrial robots check thier own accuracy and also performance besides training themselves to improve their use of AI. Cobots utilize machine vision to work securely alongside human workers.

Google utilizes AI in its data centers to enhance energy performance. AI aids to transform manufacturing by minimizing its environmental impact. Chatbot is an additional basic AI application that most of the online business portals do have, and presently, it uses augmented reality too. Chatbot takes advantage of NLP to assess text fields in studies and performs evaluations to reveal insights thereby boosting client satisfaction and effectiveness. AI in the medium domain helps in discovering brand-new drugs based on previous information and medical knowledge; it assists in reducing the cost of research and development and delivers better result and performance. Integrating Food and Drug Administration (FDA) information, AI helps in transform medicine discovery by locating medicines in the market which are FDA approved or rejected.

> AI chatbots to improve providing firm details on their official webpage. Using AI will absolutely be important for SMEs; nevertheless, business proprietor will absolutely need being future targeted and additionally prepared to touch the next edge with the most up to date modern innovation.
>
> *(Adam et al., 2021)*

AI assists manufacturing organizations, NPD team to make products utilizing generative design approach. Technology is transforming the means product designers design the smart products of the future. A designer inputs the design objective right into generative design algorithms, which explore all the possible permutations of an option as well as generate design alternatives. Ultimately, it uses machine learning to check each iteration and also surpass it. Application of this innovation aids in discovering distinct means to reimagine parts across industrial sectors while building products. In words of a product design engineer, "CAD applications that autonomously produce a number of design options provided a set number of constraints support, freeing product designers for other tasks, and on conclusion, the designer is provided the option to choose which generated design they want to check out more completely."

As modern technology grows, AI is more easily available for manufacturing industries, which are eager to accept new innovations. AI adoption in manufacturing industies enables them to make quick, data-driven choices, enhance production procedures, minimize functional prices and besides improve presentation to the clients. This does not suggest that production will be controlled by the AI powered machines; AI powered machines exist merely to assist human work and can never substitute human intelligence or man's innate ability to adapt to unexpected transformation with the arrival of Industry 4.0 and later Industry 5.0.

4D Printing

4D printing emerging technology that incorporates 3D printing strategies with high-level product science, engineering and software program. It makes use of liquid crystal elastomers, shape-memory polymers and hydrogel, which are capable of modifying the physical and thermomechanical shapes in a programmable method based on customer input or independent picking up. The technology is still in research. Basically, 4D printing is an improvement on 3D printing wherein the printed items transform shape post-production. A trigger might be water, warmth, wind and other types of energy. Lowered expenses, enhanced software application designs and variety of printable materials have led to the development of a new modern technology called 4D printing.

> NASA's Jet propulsion Lab has actually developed associate degree flexible steel material that could be used for big antennas, to protect a ballistic capsule from meteorites, in cosmonaut spacesuits, as well for capturing things on the surface.
>
> *(Landau, 2017)*

4D printing is strongly influenced by the concept of self-assembly, a concept commonly utilized in nanotechnology. The key distinction is that 4D printed things change their form with time as soon as they are published, whereas 3D published products retain the same, fixed form, and in regard to material, it utilizes specially designed "smart" materials that have several commercial properties that can be transformed by outside triggers. Industry 5.0 inspires 4D printing because it will certainly assist in focussing on product designing, as opposed to the manufacturing process. The freedom of designing will certainly lead to the development of products, which are extra bespoke and unique.

CHALLENGES OF INDUSTRY 5.0

Manufacturers are still actively developing approaches for interconnecting new modern technologies to enhance effectiveness above and beyond performance. The directing principle behind Industry 4.0, the following stage of industrialization, is already upon lots of manufacturers across the globe. As a result of a higher level of automation in the industries, the existing business strategy and organization models have to be modified and tailored to meet the demands of Industry 5.0. As a result of mass customization, the company method will be concentrating much more on client-centric operations. Organization techniques in Industry 5.0 demand higher degree of dynamism to survive the competition due to differential customer preferences.

Client's bias changes with time, and also, it is tough to transform business methods and also organization designs often. Smart manufacturing system along with smart materials requires higher autonomy and also sociality capabilities as crucial factors of self-organized systems. Working along with robots sounds fantastic, but employees will have to find out how to work together with a smart machine. Beyond the soft skills required, technological skills will certainly also be an issue. Industry 5.0 rollout is rigid due to lack of freedom in the present systems such as incorporated choice making.

BENEFITS OF INDUSTRY 5.0

Client satisfaction, as one of the major aspects of industry development, ensures the positioning of products. Customers can mention their preferences in the design phase and the production line can incorporate the specified preferences, with no costs included. Industry 5.0 provides the NPD team the capacity to automate manufacturing, to obtain the real-time data for the analysis and also use that data in the design process. The digital transformation of Industry 4.0 means smart manufacturing facilities with devices linked to the Internet. Manufacturing enterprises produce, accumulate and also analyze information throughout the supply chain to determine methods to drive quality improvement, process optimization, expense reduction and also compliance. Industry 5.0 will incorporate the precision along with the rate of industrial automation with the crucial thinking of human intelligence. It adds durable and lasting policies, where even a minimal generation of waste become important, cross-cutting procedures and makes the organization extra effective and environment-friendly.

SUMMARY

Technology-driven transformation requires the appropriate organization culture and the management executives to function appropriately. Modern technology alone is not enough to drive the transformation; business leaders need to engage with their workers to encourage understanding and acclimatization. Manufacturing industries that take care to foster the appropriate culture around these new technologies will be the ones with a competitive advantage, improving their existing business models, developing new possibilities, while drawing in and also retaining brand-new skill. Strategic investments continue to be vital for every manufacturing organization's ongoing development. Even if different aggregating techniques in varied operations can be come complex, the process aids manufacturers see high returns in an increasingly competitive environment. This is really a future that provides value to the manufacturing. A key aspect in improving business performance is possesing the most efficient processes and the most effective people, focusing on our client's outcomes and using cutting-edge technology to identify areas for improvement to leverage engineering process effectiveness, through manufacturing effectiveness, across different levels of the enterprises.

BIBLIOGRAPHY

Adam, M., M. Wessel and A. Benlian. "AI-based chatbots in customer service and their effects on user compliance." *Electron Markets* 31 (2021): 427–445. https://doi.org/10.1007/s12525-020-00414-7.

Ashton, T. S. 1962. *(Thomas Southcliffe). The Industrial Revolution*, 1760–1830. London, New York: Oxford University Press.

Brettel, M., N. Friederichsen, M. Keller and M. Rosenberg. "How virtualization, decentralization and network building change the manufacturing landscape: An industry 4.0 perspective." *FormaMente 12* (2017): 37–44.

Elangovan, Uthayan. 2019. *Smart Automation to Smart Manufacturing: Industrial Internet of Things*. New York: Momentum Press.

Famous Scientists. n.d. Quick Guide to James Watt's Inventions and Discoveries. https://www.famousscientists.org/james-watt/.

Giedion, S. 1948. *Mechanization Takes Command*. New York: W.W. Norton.

Gonzalez, Carlos M. 2021. Is 3D Printing the Future of Manufacturing? https://www.asme.org/topics-resources/content/is-3d-printing-the-future-of-manufacturing.

Kolberg, D. and D. Zühlke. "Lean automation enabled by industry 4.0 technologies." *IFAC-PapersOnLine 48*, no. 3 (2015): 1870–1875.

Landau, Elizabeth. 2017. 'Space Fabric' Links Fashion and Engineering. https://www.jpl.nasa.gov/news/space-fabric-links-fashion-and-engineering.

Laurean, Bogdan. "Programming and Controlling of RPP Robot by Using a PLC." *Annals of the Oradea University. Fascicle of Management and Technological Engineering* XIX, no. IX (2010): 2010/1. https://doi.org/10.15660/AUOFMTE.2010-1.1764.

Lee, J., H.-A. Kao and S. Yang. "Service innovation and smart analytics for industry 4.0 and big data environment." *Procedia Cirp* 16 (2014): 3–8.

Microsoft. n.d. A New Reality for Manufacturing. https://www.microsoft.com/en-us/hololens/industry-manufacturing.

Mowery, D. and N. Rosenberg 1989. *Technology and the Pursuit of Economic Growth*. Cambridge: Cambridge University Press.

Outman, J. L. 2003. *1946- and Elisabeth M. Outman, Industrial Revolution*. Detroit: UXL.

Schwab, K. 2017. *The Fourth Industrial Revolution*. London, England: Portfolio Penguin.

Stanley, Jevons, H. "The Second Industrial Revolution." *The Economic Journal* 41, no. 161 (1931): 1–18. https://doi.org/10.2307/2224131.

Wang, S., J. Wan, D. Li and C. Zhang. "Implementing smart factory of industries 4.0: An outlook." *International Journal of Distributed Sensor Networks* 2016. Article ID 3159805. https://journals.sagepub.com/doi/10.1155/2016/3159805.

Wheeler, Andrew. 2015. Lights-Out Manufacturing: Future Fantasy or Good Business? https://redshift.autodesk.com/lights-out-manufacturing/.

Witkowski, K. "Internet of things, big data, industry 4.0–innovative solutions in logistics and supply chains management." *Procedia Engineering* 182 (2017): 763–769.

4 Transformation in Automotive Sector

Worldwide competitors, rapidly changing technologies, lowered product life cycle, price reduction, high-quality products and demanding end customers are several of the elements that have actually made manufacturing enterprises to search for new techniques for establishing new product development. One of the most significant inventions the world has ever witnessed is the automotive sector. Manufacturing sector and automotive sector were strongly linked throughout the twentieth century, and such ties will most likely remain pertinent in the future too. The automotive sector includes not just vehicle manufacturing, but also components and parts that are required to assemble a single vehicle; besides, a number of sectors are associated with their manufacturing, such as steel, glass, plastic, rubber, fabric and electronic devices. Currently, automotive sector is undergoing massive technological innovation, from appearance to speed and advanced capabilities; the automobile of today is smart and extra power effective besides continuously evolving.

The path breaking technological development made in the automotive market was the introduction of full-blown automation, a process combining precision, standardization, interchangeability, synchronization, as well as connection. The evolving digital transformation of product lifecycle expectations and the need for new cutting-edge solutions will certainly influence all elements of the industry. The industry is reaching an inflection factor in which electronics and software application will displace mechanical equipment as the most vital components. The influences of automotive, technical and market fads are not restricted to the design and manufacturing alone. This will have significant repercussions in the automotive sector; they drive essential modifications in business models and functional frameworks as significant industry transformers.

• Automotive market is among the leaders of the fourth industrial revolution; however, there is a huge void between the original equipment manufacturer and discrete manufacturers comprises SMEs, a lot more prevalent of Tier 1 suppliers and Tier 2 suppliers. Quick technological growths resulting in improvements in design and manufacturing, boosts in electronic driving systems, altering customer choices, expanding concern about sustainability and regulative stress and measures to transform the frameworks and developments in batteries have actually brought about substantial price reductions opening up great deals of possibilities for electric vehicles (EV) manufacturing in addition to its facilities.

Automobiles have and always will be an important part of human daily living. Tracking the advancement of the vehicle and its parts from its basic phase to its present degree of luxury raises some inquiries, such as: What will the future hold for passenger vehicles? What will be the capabilities of the automotive industry with

DOI: 10.1201/9781003190677-4

industrial transformation? Will factories in the future have the ability to operate on a combination of machine intelligence and human intelligence? Having such sophisticated modern technology, how will the growth be? Let us go further to touch base few areas.

PROCESS REVOLUTIONS

The growth of automotive products has certain properties such as managing complexity, traceability, awareness of the standing of information, trust in understanding, dependability, vast use distributors, severe competitors, high development price, long lead times, high degree of expertise strength, quickly changing technologies and also the inherent dangers. The emphasis is on creating an item that meets the standards of a premium client, so that the item can end up being a success on the market. SMEs/OEMs have to attain success through carefully implemented new product development process; it is a vital procedure for the success and survival of companies in the automotive sectors. NPD process includes all the tasks from the approval of a suggestion or a concept for a new product, to the realization of the product during the manufacturing stage and its introduction into the marketplace. Generally, the NPD procedure comprises different stages till the product is released such as planning, product and process design and development along with procedure endorsement.

To make NPD effective, there needs to be a synchronization between the production, engineering, research and development, advertising and marketing, finance and purchasing departments. The difficulty lies in developing a procedure for successful product innovation, where new product jobs can move quickly and also effectively and efficiently from the concept stage to a successful launch the process of development from the preliminary idea to the final authorization of the finished design spans a period of years in which the design group jointly produces the end product information. Quality function deployment (QFD) is utilized to convert client demands into product and process design needs and identify the technical demands that need urgent improvement, as it entails not just the customers but additionally the competitors.

The main challenge of the automotive industry for both the component supplier and OEM is concentrating on the quality per cost ratio of the product manufactured. Customer expectations are constantly transforming; so, automotive manufacturers need to pursue continuous renovation, to make sure that errors can be avoided. So, quality assurance makes certain that each item leaving the factory is of the highest quality meeting the consumer expectation. One international quality standard that is endorsed by many nations and automotive manufacturers is the IATF 16949 Technical Specification. IATF 16949 helps manufacturers to improve their effectiveness, performance efficiency, flexibility and safety throughout the supply chain. It provides a framework for accomplishing the finest quality practice by an automotive manufacturer, from design (product and process) till the production of the end product delivered to the customer. Quality tools that any automotive center can make use of to boost their quality assurance strategy that supports IATF 16949 are advanced product quality planning (APQP), failure mode and effects analysis (FMEA), SPC, production part approval process (PPAP) and measurement system analysis (MSA). Client positioning, creative thinking and also development are important variables that affect the product growth process and are closely interconnected with high quality in the NPD process.

There are a few quality methodologies followed in different regions across the globe, which are as follows.

Six Sigma

Six Sigma methodology is a business performance enhancement method (instituted by Motorola), which intends to reduce the variety of errors as well as defects to as low as feasible per million opportunities. It contains Define–Measure–Analyze–Improve–Control (DMAIC) and Define–Measure–Analyze–Design–Verify (DMADV) methods that eliminate defects from a process. Six Sigma DMAIC approach in an automotive sector supplies a framework to determine, measure and remove resources of variation in an operational procedure, as well as optimize the problem variables, improve sustainable efficiency via process return with well-performed control strategies. Six Sigma DMADV method in an automotive industry provides a framework to create new a defect-free product and process to meet critical to quality (CTQ) aspects that will certainly assure client satisfaction. Design process is one of the costliest and also time-consuming phases; countless modifications following late detection of product design errors are significant troubles that one comes across throughout the automotive component and automobile perception phase. So, when dealing with physical mock-ups, frequent reverting to previous decisions and limitless modifications the overall task expense is significantly elevated. Identify, Design, Optimize and Verify (IDOV) is a phase process utilized by design by the Six Sigma (DFSS) team for designing products and services to meet Six Sigma standards.

Toyota Production System

Waste can materialize as excess supply, supplementary handling actions, and malfunctioning items, to name a few circumstances. All these "waste" elements combine with each other to produce more waste, ultimately influencing the administration of the automotive enterprise itself. The Toyota Production System (TPS) was developed by Toyota Motor corporation based on two concepts: "jidoka" (automation with a human touch), as when an issue takes place, the equipment stops instantly, avoiding defective products from being produced; and the JIT principle, in which each procedure produces just what is required for the following process in a continual flow. Through the repetition of TPS process, shop floor machines end up being less complex and more economical, as maintenance becomes less time-consuming and much less pricey, making it possible for the development of basic, slim, versatile lines that are adaptable to fluctuations in manufacturing quantity.

Lean Manufacturing

Lean manufacturing is a technique that enhances the process with continuous improvement (kaizen) as well as elimination of waste. Lean principles have actually revolutionized the automotive market, permitting them to reduce costs and improve their performance. Lean manufacturing has migrated to a more comprehensive area of implementation called lean management. It is a common process administration viewpoint derived mainly from the TPS. Lean manufacturing deals with a tried and tested approach to get rid of non-value-added activities and also waste from the enterprise process. It concentrates on lessening the human initiative, production room, financial investment in shop floor machines and reducing the engineering time to develop a new product. Value stream mapping (VSM) enables manufacturers to develop a strong application strategy that will maximize the available resources. For a lean manufacturing journey, VSM functions as the launching pad to start determining the best ways to improve their process. The goal of VSM is to recognize, demonstrate and lower waste at the same time, highlighting the chances for renovation that will significantly affect the total manufacturing system.

> Truth for SMEs is that, as a result of the resource constraints, it is particularly crucial that management executives are the ones to have extensive understanding of the lean to enjoy its benefits. Lean calls for dedication and also the involvement of everyone within the SME. It is typically extremely simple to function around the principles of lean to hit short-term targets.
>
> *(Alkhoraif et al., 2018)*

SMEs performing lean usually have a framework and easy systems, which promote versatility to continuously evolve as well as disseminate information. Lean iceberg models discuss that the execution of lean tools and procedures requires unnoticeable components of determined positioning, management and additionally involvement. The stress to fulfill the demand must be carefully maintained by retaining and even improving the quality. This is where lean manufacturing concepts come to play. With the implemetation of Industry 4.0 and IIoT, lean goals will be completed a lot more immediately. Lean principles are incorporated with less side innovations that make it feasible for constant, real-time surveillance, quicker decision-making, boosted effectiveness, along with the leanest procedures feasible. Lean is a journey not a final boundary.

WORLD-CLASS MANUFACTURING

World-class manufacturing (WCM) is the ideology of being the best, the fastest and also the cheapest manufacturer of a product and service. Fiat Group specifies WCM as a structured over and above integrated manufacturing system that includes all the procedures of the factory, the safety atmosphere, from upkeep to logistics and similarly high quality. WCM suggests consistent improvement of products, process and solution to remain an industry leader and also supply the most effective choice for clients, regardless of where they are in the procedure. WCM calls for all decisions to be made based on unbiased measured information and its analysis. It aligns people, process and also innovation capacities to develop a culture of continual enhancement, targeting zero losses, client cases, quality flaws, device malfunction as well as accidents.

Quality Control performed across the supply chain as modifications in manufacturing procedures, consumer demands and disruptive trends all affect the automotive supply chain network for resources, parts and finished components. Vendor development is indeed an essential role, streamlining the flow of components in between suppliers and manufacturers in automotive NPD and NPI processes. Now, there will/ might be questions that arise in the minds of SMEs: Which above-mentioned method should be used to improve the process? SMEs have much less experience with business enhancement approaches; it is far better to go ahead with the stepping stone approach. Process designers evaluate and create procedures to increase performance and also range their business services. The decision to choose among the techniques does not need to be puristic. Consider the requirements from customer together with the changing demands as the baseline, which is the starting point to convert exact customer demands into products, by maintaining the quality, throughout the product development process. A complete and excellent execution of one technique is not the goal. Be versatile and continue to be to think out of the box to carry out an improvement technique, if necessity demands. It is only the outcome that matters, which is of far more worth to the service and consumers. With the industrial transformation and the arrival of innovative manufacturers, the early idea of quality transformation was previously considered unpredictable as well as largely identified by the skills of individuals. The underlying success of the continuous improvement technique lies with leadership support, involvement, tactical focus and also execution.

Technical or industrial technology is used to describe a new breakthrough in a procedure or a manufacturing method or a unique product, and it is utilized extensively by economists. The products are configurable with set of services to additionally improve the product value and its usage. The challenge in designing a system of automation components of lasting worth lies in understanding the possible needs of the future. A specific synchronicity is called for in order to align product development with automation through the production operation. Enterprises requires to take on an electronic improvement effort with the goal of tying in all the relevant silo systems to develop a single collection of relevant information that flows via an ecosystem of seamless data connection, obtainable to all service partners, both internal and external, called the digital thread. It aids in upstream and downstream work with the exact same product definition information that is trustworthy and workable, consequently, delivering high-quality products avoiding several interpretations of the same information enabling interaction of engineering modifications, making quicker decisions and also executing them easily as it is readily available to all cross-functional teams of NPD/NPI in the item supply chain. Increased collaborations between product engineers and production engineers assist in designing manufacturing procedures. By linking smart items, manufacturers can collect feedback from the product's field performance and use; thus, boosting product design can produce brand-new organization chances for customers.

BUSINESS NECESSITY FOR PROCESS TRANSFORMATION

In the industrial world, technical innovation is rapidly changing, and the future is not certain; correspondingly, it will be important for automotive businesses to swiftly maximize brand-new possibilities and benefit from this disruptive development. Process transformation is the buzzword being talked about in every conference room of SME (Tier 1 or Tier 2 suppliers) and OEM. Automotive suppliers and OEMs are called for to prepare for a major change throughout the whole value chain, as a result of process transformation in the production as well as in the supply chain process. Process change is to attain essential objectives such as using technology for much better service outcomes and optimize company processes with electronic innovation. There is no single dimension that fits every supplier and OEM for effective process transformation. Every supplier and OEM of automotive industry is different, with their very own method of doing things and abilities. It is understanding what those strengths are and how to use them that makes the difference between the actual transformation and the unrealized aspiration.

Any technological and process development is bound to have a solid effect on the established standards starting with product development, manufacturing, distribution and the everyday use by the end consumer. Process transformation that functions well for an automotive business can bring substantial, favorable adjustments in operations, waste reduction, expense reduction, high-quality products and also go-to-market strategies. It usually involves an assessment of the actions called for to accomplish a particular goal in an initiative to remove non-value-added steps and automate as many actions as feasible. Enhancing a consumer's journey requires the automotive industry to recognize every modern technology, process, ability along with the transition required to provide a terrific experience. Sustaining in the digital, smart connected age requires predictive intelligence and ideal timing to respond to the tranformations in the industrial economy.

First and foremost is to focus on the existing abilities that have a direct effect on the success of the process transformation. The aim must be to make the necessary changes to the procedures, individualsand technology to align the business with the organization's strategy and vision. Almost, it is the procedure of end-to-end customer experience optimization, functional flexibility and technology development that are key drivers of organization digital transformation model which opens up new income and value. Automation has proven to be an efficient transformation tool to boost productivity in production. The incredible progress in robotics, artificial intelligence and machine learning, and so on will assist SMEs to OEMs of automotive industry to become much more affordable with design and manufacturing cycle time reduction. With several new modern technology innovations and market shifts striking the industry at the same time, automotive discrete suppliers are facing one of the most difficult settings of the last century. Are automotive components manufacturers or OEMs prepared for today's auto industry's evolution patterns? If so, process transformation should be considered as a long-term journey that starts with a few concerns that need to be clarified before commencing the initiative.

PROCESS TRANSFORMATION REVOLUTIONS

Digitized technologies in the production are changing every department in the automotive manufacturing, from product development to manufacturing to sales through services. The fast pace of process change is transforming the component hardware-driven automobile sector to a software program and solutions-focused sector, accelerated by the customer's advancing expectations and needs for new digital innovative services in the era of Industry 4.0. The quick growth of robots, data analytics and also the rise in the use of the IIoT allowed digital connection by linking the firm's management and process with the customers. Awareness of technological innovation and its effect on the business model needs to be high; therefore, a great deal of effort and time has to be invested in bringing this bent on the brand-new environment, not simply from a technological perspective but also regarding the security process effects of the digital and ingenious modern technologies. To attain functional excellence and strategic improvements, automotive leaders need to touch base transforming different service procedures within their venture by looking past modern technologies. Few areas of process transformation in the automotive sector are discussed here, which will/can assist SMEs and OEMs.

High competition in the automotive industry pressures manufacturing enterprises to invest in better equipment and smarter options to boost the high quality of the new product without jeopardizing the timing. The mainstay of modern technology that develops the base for automotive sector is high-performance computing that comprises of CAx, PLM, ERP and MES.

Implementing PLM for SMEs

Products are the lifelines of all automotive sectors in addition to becoming more advanced and smarter. CAD data give product details for all product manufacturing SMEs and OEMs. To maintain information efficiently, these CAD details need to be shown to all the CFT of the product design and development. Originally founded as engineering data management, these systems managed CAD data and subsequently developed into PDM systems taking care of CAD documents, bills of materials, variations and also revision control. PLM manages all aspects of the product life cycle, from concept design to product retirement, collaborating with extended enterprises across the globe.

Business Challenge

SMEs face the same challenges in handling 3D CAD data that the OEMs manage. Engineers and designers attempt multiple product design alternatives, to find the best solution. To manage the complexity, they need to track what they designed yesterday and the week before, in addition to what they wish to retain, replace or review and approve progression; thus, the process gets extremely unpleasant for the product development team.

Precondition

CAD application for modeling product design comprises of design data as well the details about the product such as part number, type of the part, customer and revision as per the customer requirement along with the NPD/NPI program.

Approach

Essential components of the PDM innovation and execution process need few dedications as it involves the entire CFT within the enterprise, the vision and the management strategies to roll out. Begin with, two crucial functions are configuration management and process management. The major task for the configuration management is to keep track of the right set of documents for each version of an item. Process management is used to automate numerous procedures in the enterprise. Leveraging a protected vault extends the access to the 3D CAD environment along with its associated data, for all the participants from engineering to production. This makes it

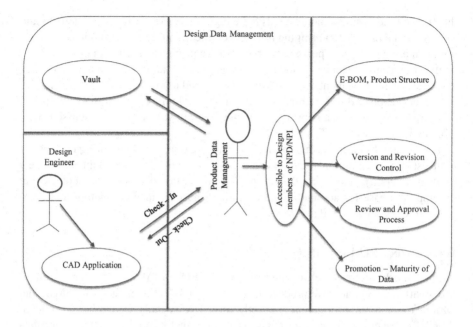

FIGURE 4.1   Implementing PDM: Foundation of PLM.

possible for every person associated with the tasks to share details and team up on designs, while automatically safeguarding the product copyright with the automated variation and revision control.

Result

PDM as the core of PLM provides complete configuration control of the product data from all phases of development, from initial idea, through design, development and manufacturing. It provides the product development team the path to access all product-related details on the various restraints and also demands at the different phases of the product life cycle. PDM and PLM are currently driven primarily by cloud computing technology. Cloud is a perfect standard to share data. It speeds up the moment of implementation, gives adaptability and also decreases the overall cost of ownership. Cloud PLM boosts advancement and flexibility across the community, making it possible for the extended enterprise. SMEs are anticipated to constantly produce faster besides reinventing products, while boosting sustainability in this Industry 4.0. It also decreases both the cost and application of PLM as well as paves the path to digitalization. SMEs require to begin their process journey with PDM and then go on to PLM supporting the NPD process.

IMPLEMENTING PROCESS CONTROL SYSTEM (PLC, SCADA) FOR SMEs

Automation has taken over the assembly line of automotive manufacturing, which includes four process, namely welding, stamping, painting and setting up. A service technician remains in place to monitor the PLC to discover especially what is malfunctioning and proceed to take the required action. PLCs have contribute a lot to process transformation in the production shop floor and correspondingly play a major role in industrial companies, especially for small and medium sized manufacturers planning to adopt Industry 3.0. MES integrates multiple control systems that supply visual monitoring applications; one of which is the application of SCADA to collect real-time information that can effectively regulate and keep an eye on industrial machines along with the manufacturing processes. Also, it forms the basis process drive of IIoT.

Business Challenge

A significant challenge encountered by a majority of the SMEs and OEMs is the difficulty in connecting numerous shop floor machines along with machine tools that do specialized actions, share information among different devices and equipment in real-time and assembling it into a legible and actionable task influencing elements of the automotive production process. Process control automation is rather remarkable to look at. Many of today's products are made with the aid of a closed-loop signal chain, with less intervention from the operators. The manufacturing floor requires accuracy and a limited number of failings, so the manufacturing process needs to be frequently measured and also regulated.

Precondition

Hardware – PLC, Software – SCADA, Human–machine interface (HMI), DCS, MES, Shop floor machines, Sensors, Actuators.

Approach

PLC's features are separated right into three major classifications as inputs, outcomes as well as the CPU. Innovation in industrial automation still processes with some types of hand-operated controls, which does not always guarantee optimal performance. By using control tools, it is possible to optimize the procedures, provide a secure and reliable operation with data offered much more quickly. PLC is an equipment that gets information from connected sensors and input devices, processes the data, and triggers outputs based on pre-programmed specifications. It records information from the shop flooring by monitoring inputs that devices and machines are linked to and by utilizing software program SCADA. The production line operator monitor and regulate the PLC and record data, even from remote locations. Collaborating SCADA, MES and HMI systems, together with an enterprise-wide solution, allows manufacturers to see and control information on a PLC.

Result

The targets will be achievable by having PLC and SCADA and by establishing significant rigid delivery routines, thereby increasing production, effectivity as well as performance through the procedure of information technology systems monitoring and industrial control devices.

HEAT TREATMENT PROCESS

While talking about the manufacturing process of the automotive components, material plays a vital role, and how it is heat treated for strength and stability is a must to go through. Despite the arrival of electric cars, heat treatment of vehicle components remains to be existed. It plays a crucial role in product development and sustaining new technologies. Advancement in heat treatment process across various industrial sectors takes care of preservation of power inputs, concerns about environment, utilizing standard materials in modern non-traditional ways and executing value-added procedures to upgrade existing materials to meet a lot more rigid guidelines, safety of human labours and customer demands. The introduction of lightweight aluminum alloy in the automobile body and structural parts has caused extra obstacles for thermal processes and equipment that improve the component strength and ductility characteristics. Correct control of distortion after thermal treatment of powertrain elements in the automotive market is a vital action in ensuring top notch componentsas well as to minimize hard machining procedures in order to decrease overall production costs. There is an ever-increasing product need for steel and aluminum materials with boosted mechanical and metallurgical properties to fulfill the challenges of the automobile manufacturing procedures. Below is the use case of SCADA used to improve safety, streamline and automate heat treatment processes.

Business Challenges

The automation of heat treatment process has big application in automotive industry. The majority of this process is carried out in hand-operated situations in SMEs. The person needs to monitor the home heating chamber constantly and preserve the temperature and time duration of heat applied in steel throughout the whole heat treatment process.

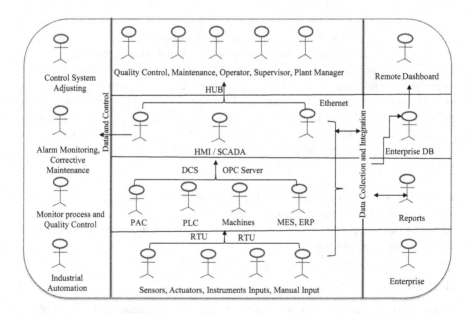

FIGURE 4.2 Shop floor process control automation.

Precondition

The important conditions for the heat treatment process are exact temperature level control, accurate structure of the atmosphere and specific timings. High volume heat treatment of material proceeds with continual heating process, and that the material is constantly moving into and out of the heat chamber without actual start and stop point at the same time.

Approach

SCADA provides heat treatment plants amazing abilities for remote operation and surveillance of the heating systems with user-friendly online devices. It helps control processes and immediately accumulates information ensure that the systems are running accurately, and every part of the process is accurately recorded. PLC helps to regulate temperature settings, speed of the fan and so on. PLC helps to regulate temperature settings, speed of the fan and so on, those at the entry and exit terminal control, which includes the electric motors and hydraulic cylinders of the handling equipment. SCADA ensures that there is power to allow different items to travel through the workstations in various cycle series, while ensuring complete integrity and compliance requirements set forth in widely used industry standards such as AMS2750 in aerospace industry and CQI-9 in automotive industry.

Result

Diagnostics are made certain exact alarms that alert the operator or maintenance worker right away to identify any anomaly and promptly secure the typical operations. PPAP can additionally be made use of to establish and record fixed heat treatment procedures besides forcing the SME resources to demonstrate process capability prior to introducing manufacturing. It ensures formal quality preparation and forces vendors to report and document any kind of process adjustments, prevents the use of non-conforming products, and reduces the possibility for guarantee claims.

PREDICTIVE MAINTENANCE IN HEAT TREATMENT PROCESS

Predictive maintenance of the heating process furnace will show whether particular maintenance activities are required, shifting from a timing-based maintenance to a condition-based maintenance. The energy consumption is dependent on power as well time. A lot of the industrial sectors are utilizing vacuum heating systems as opposed to blast furnaces as they produce much less carbon dioxide, and it is environment-friendly. IIoT assists to identify problems of any kind, be it part of the system or the whole heating system, and anticipate malfunctions, thereby determining one of the most affordable times and approach for maintenance without the loss of productivity. The integrity of devices being connected to a heat treatment device has enhanced dramatically whereby the environment can stay clear of physical collection of information by a service technician. Modernization of the heat treatment shop floor is crucial to both success and survival of the industry. Predictive maintenance technology is becoming a powerful tool for heat treatment in the automotive sector for analyzing performance and effectiveness.

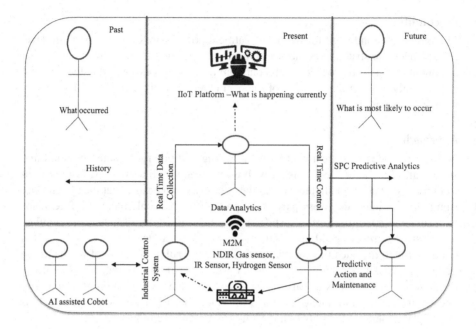

FIGURE 4.3   IIoT and cobot in heat treatment process.

ADDITIVE MANUFACTURING

Faster product advancement along with technological innovation is vital to an enterprise's success; rapid prototyping becomes one of the most essential elements of new product development. Quick fabrication of physical models making use of three-dimensional CAD data turns ingenious concepts right into effective end parts, swiftly as well as efficiently. Product layouts, and prototyping procedures of building, assessing and fine-tuning, suit all significant stages of the design process. A prototype is a preliminary variation of the end product; it is made use of to assess the design, test the technology or assess the working principle, which subsequently supplies item specifications for an actual working system. Rapid prototyping tool for automotive parts design process has now advanced in the direction of additive manufacturing (AM) or 3D printing, which is a game changer for OEM and SME automotive component manufacturers.

> Generating spare components is a classic example of 3D printing. Porsche supplies parts for its vintage and out of production designs, using 3D and Ford incorporated 3D printing into their product design and development process, it creates 3D-printed prototypes utilized for layout validation and also functional screening.
>
> *(Newsroom, 2018; Henry Ford, n.d.; Ford, n.d.)*

AM enables quick prototyping in the pre-manufacturing stage. One of the most popular ways is to confirm a model from a small promptly published information to a high detail major component ideal for efficiency validation. It reduces the cost involved and also long lead time CNC production. The parts created by the 3D printer are cheaper, and their in-house manufacturing time is much shorter. It aids designing in the automotive industry enabling product developers to attempt several choices of the very same information and iterations throughout the stages of NPD. 3D printed parts are more ergonomic and yield higher operator interaction as comments can be easily included into design models, all adding up to unmatched effectiveness levels.

VIRTUAL REALITY AND AUGMENT REALITY

Automotive vehicle and component designers interact with the vehicle prototype to better recognize communications between components and parts. Product design reviews end up being more efficient, as virtual reality (VR) helps businesses to dramatically decrease the number of models needed for the automobile promotion. Consequently, the advent of VR and AR is the most promising innovation of Industry 4.0, which has spread its arms in the automotive industry extensively. VR gives the opportunity to gather all CFT of NPD/NPI and suppliers in the exact same virtual work space; thus, it favors explanation of misunderstandings and brings about quicker and a lot more reliable decision-making and offers broad capabilities for training engineers. AR is obtaining traction quickly, driving value for the enterprise and the clients.

> AR technology in an automobile is the motorist experience improvement, can be discovered in the type of transparent display screen screens in windshield task more details concerning surrounding environment, troubles, as well offering instantaneous details on any kind of vital events, without sidetracking from driving.
>
> *(BMW UX, n.d.)*

The exceptional thing about AR is that it enables users to connect with the real world making use of technology-supported graphics. For things that can be quickly solved without an auto mechanic, operators can provide this technology for self-service.

QUALITY STANDARDS

The development of autonomous cars will change the means the automotive sector operates. Large scale adoption of this modern technology is reliant on a significant, however not impossible, change in perceptions. The difficulty for the automotive market is the critical timing of investment according to transforming attitudes. The potential to access new markets is not restricted to automobiles alone; moreover, the worldwide need for vehicle parts is perpetually increasing. The changes present distinct production obstacles to electronic manufacturing services (EMS) providers, who serve automotive OEMs and their suppliers.

Considering the enhanced consumer satisfaction, safer products, improved performances and also a raised bottom line, it is easy to see why the value of a quality assurance system is growing in the automotive sector. Security is of highest importance, and, a top-quality monitoring system is a crucial means for vehicles along with their parts to pass safety examinations and standards. Automotive products need to fulfill conformity besides functional quality, making sure that industry ideal practices are being followed across the enterprises. So, one such technique is IATF 16949, a globally acknowledged high-quality management standard, which provides a framework for accomplishing the finest technique in an organization, touching all areas from design through production of the end item and its components that enter into the automotive supply chain. SMEs that lack the resources to address these progressing demands might require to utilize the quality assurance capabilities of their bigger business partners, consisting of distributors and examination subcontractors. It means that SMEs need to have their community partners provide specific contractual guarantees to the customers. SMEs also need to take actions with their mostly contracted out supply chain to ensure continuity of supply, second-source needs and also conformity with hazardous substances and ethical sourcing laws. OEMs have been understood to involve with SMEs while at the same time developing their very own remedies, often learning from the SMEs at the same time.

Numerous complex products have flooded the automobile market as most automobiles are packed with every sensor, consumer electronic and infomercial option feasible, for instance, the expanding number of chauffeur assistance systems required to counter all the distractions created by the various other innovations in the cabin. Product and process development in the automotive industry has actually been progressing impressively. An automotive supplier to OEMs accomplishing these market requirements shows that it can deliver higher product top quality at lower failure rates. One such industry standard is automotive software performance improvement and capability determination (ASPICE). It offers the framework for specifying, executing and assessing the process required for system growth focused on software application and also system components in the automotive industry. During the supplier option, an OEM can utilize the ASPICE structure to examine the ability and quality of the distributor. On the other hand, ASPICE can prove to be an ideal structure for the suppliers to take their existing high quality a few notches greater. This includes the support, that Tier 1 and Tier 2 suppliers can offer their products, manifesting how reliable and consistent their interior processes are.

The progression from the third industrial transformation to the fourth followed by the from fifth industrial transformation in the direction of digitization consists of automation, robotics, cobots, artificial intelligence, machine learning system. It is crucial for the automotive SMEs and OEMs to rapidly adapt to these radical transformations in the market to top the competitive landscape. Organizations need to be developed to be robust and also to deal well with the increasingly fast pace of process transformation.

BUSINESS CHALLENGES IN PROCESS TRANSFORMATION

Process transformation, automation and the introduction of new business models have changed the industrial sector in the last few years, and companies need to adapt to the new ecosystem. Patterns such as enhanced connectivity, ecological guidelines, IIoT, wireless services and client assumptions drive investments into process improvement in the automotive industry. Process transformation efforts loop technology-driven fads along with consumer demands to stay affordable in this smart connected manufacturing world. Smart connected supply chains drive costs down, involve the consumers better and also collect user data for a much better service. A few of the challenges are to set manufacturers 'cost planning and preparation, emission planning, accountancy, inventory monitoring besides manufacturing preparation. In the context of process transformation, these tasks become increasingly more complicated due to the use of large amounts of information.

The common challenges of small and medium enterprises are lowered client buying power, limitations on communication, lacks of resources, cancellation of orders, cash flow problems and also supply chain disturbance. To embrace the industrial transformations (Industry 3.0 to 4.0 through 5.0), small and medium sized enterprises should respond quickly and establish an overarching digitization strategy so as not to be overwhelmed by the plethora of possibilities and need to become extra nimble, quicker and also bolder. SMEs need some positioning along with a general understanding of the process transformation initially. Few SME's that are not familiar with technology struggle to understand what to digitalize, which technology to utilize, exactly how to focus on objectives and which business changes (e.g., skills and also functions) are required. It does not always have to be big and costly applications that bring SMEs to achieve the goal. A lot of these buisness applications are cloud-based and provides fourteen days to one month of trial version to play around, which gives the sufficient for the SMEs to decide whether the corresponding application fits their business process or not. The sooner the SMEs address the above challenges the quicker they can reap the benefits and to position themselves better than their competitors.

OEMs or large enterprises have a myriad of siloed systems including numerous scraps of information regarding consumer interactions, but no clear way to pull them together. One such means is to follow the path to success via the examination and discover technique, where new functions are being regularly added, assesed, modified and trimmed, based upon individual comments and user data. The early identification of emerging markets or technological gaps or the boosted visibility of rivals motivate the enterprises to respond swiftly. They can also establishing processes designed to generate profiles of prospective concepts for the future state of their transformation journey. Dedication and an adaptive mindset are called for in all levels within internal divisions. Future development and vigor call for equivalent focus to various metrics, consisting of client contentment, partnership growth, time-to-market tracks for brand-new products as well as inner adjustment matters. One effective way is to take a step back every couple of months and also assess how businesses are profitable in this competitive market.

Overall, automotive manufacturers need to learn from modern technology trend-setters and begin to digitize their whole manufacturing procedure. Some other pertinent challenges faced by the automobile industry are safety and security, conformity, meeting customer requests and digital expectations, managing large data, working with brand-new companion's process improvement professional and developing value chain impacting every facet of the automotive industry. OEMs are leading the adoption of process transformation and ingenious innovations, whereas few SMEs are at a nascent phase. Financial investment in automotive procedure change and enterprise will need to stay focused on one of the most useful use cases with the highest possible ROI. Any kind of transformation needs to start by basically analyzing and studying the attempted and evaluated business designs, procedures and organizational structures. Based upon a convincing vision and a service technique that is derived from that, the process transformation must then address an extensive strategy, with sustainability as an essential component of the process transformation, and a strategy that guarantees that innovation is being executed in such a way that it supports business purposes.

BRIDGING PROCESS REVOLUTION TO
PROCESS TRANSFORMATION

Industrial transformation's true worth lies in ultimately shutting the void between production and engineering. When products are updated, everything required for manufacturing, from the production illustrations to the expense of materials, can stream per plant. Having live manufacturing information come directly from the design engineers will additionally cut short a great deal of the iteration and model validation time. From product concept to end of life, the emphasis of SME and OEM is to take care of product data securely, decision making and forecasts effectively.

The safety of the work environment can be enhanced by permitting a cobot to manage hazardous jobs. Cobots with electronic vision camera use simple integration for testing and quality check, ensuring the high quality of the end products identifying defective components prior to they being sent out to the last assembly. Cobots are progressively backed by artificial intelligence and machine learning technology that let them do advanced tasks along with sharing the lessons discovered. Manufacturing automation does not just include introducing brand-new modern technologies; also, the existing automation system can be upgraded to be made smaller and getting new controls becomes a lot more effective, efficient as well as attainable. Advancement of smart products that contain useful components designed to react to specific external environment along with the accessibility of 3D printing, set the stage for 4D printing. 4D printing technology in the automobile industry can bring about finishing that adjusts to nature changing conditions. 4D printing innovation incorporated with cobots enable them to print tailored reuse of automotive components.

With the help of AI, OEMs and SMEs can automate processes, equipment, devices, labor demands, forecast needs, enhance inventories, logistics and monitoring. The adoption of artificial intelligence innovation by the automobile industry has created numerous partnerships between conventional component manufacturers, technology giants and niche startups besides service providers. For OEMs, the efforts and expenditures involved in embracing AI are worth it, and they are ready to take the economic dangers. Suppliers can make use of AI-driven systems to create routines, manage processes, allow robots to function securely alongside people (to an extent) in the production line and also recognize issues in components. AI can additionally function as an enabler to automate, accelerate and enhance the accuracy of the product design and development processes. AI-powered hardware can visually inspect and provide exceptional quality control on numerous items, such as machined parts, painted vehicle bodies and distinctive steel surfaces. AI-powered innovation has actually been utilized in an automated guided vehicle (AGV). Counting on artificial intelligence without any human aid, AGVs recognize the crammed objects, readjust their routes and also supply materials to numerous parts of automobile plants. Even though difficulties still exist, such as the complexity of software growth, conformities and laws, various fields within the automotive sector are already leveraging AI technology on top of seeing enhanced effectiveness and also optimization of procedures. The largest drivers of change and development originate from using electronics, information and communication technology in automobiles. Suppliers play substantial roles in the design of systems, controlling of supply chains for manufacturing and also assembling. The innovation of Industry 4.0 are installed in cars, and the relationship between technology firms and automotive market manufacturers is additionally going to undergo significant changeover.

SMEs are the foundation of industrial economy advancement. The most critical problems for SMEs are sustaining the high-quality and constantly striving to improve, the extent and range. A few of the major variables responsible for this are substantial hand-operated treatments in processes, interrupted flow of information and also the absence of experienced manpower. The development of the automobile sector has actually assisted in the development of a big environment of SMEs catering to the automotive sector, and their standards are lifted and quality increased. The requirement of the hour is for SMEs to take the lead in adopting brand-new technologies and making them an indispensable part of their organization approaches. Those that succeed in comprehending the power of digitalization and exploiting it across their businesses will lead this new age of growth.

Leveraging Industry 4.0 technology for efficient real-time surveillance and analysis is the real need in the smart connected manufacturing environment. Manufacturing execution systems used in production, to track and also record the transformation of basic materials to end items integrated with robotic automation along with machine-to-machine (M2M) communications and enable real-time data tracking and analysis abilities that include more agility to automobile production. Tapping real-time information created from tasks on the shop floor, top quality and manufacturing facility personnel can execute root cause analysis and make procedure adjustments really promptly. Applying top quality methods and Industry 4.0 innovation facilitates quicker, real-time interaction and ensures constant international operations that equate into enhanced regulative compliance. Providing high-quality and absolutely no problem products is vital not only from the perceptive of quality control but also from the safety point of view for automotive OEMs and SME; this aslo translates right into a positive experience, the supreme standard for consumer complete satisfaction.

With the arrival of Industry 5.0, production process within an automobile industry could be considerably transformed with AI to ensure that human laborers are no longer required to do the very same tasks. Likewise, manufacturers explore the use of exoskeleton wearable industrial robots to protect human employees, making them a great deal more powerful while keeping their mobility at maximum. Robotics and cobots along with AI procedures eventually change the requirement for low-skill workers, which naturally has the potential to retrain those employees for greater level tasks. As manufacturing enterprises use IIoT in the form of advanced software application that analyzes vital data, automotive manufacturers can to work toward lean production purposes, thereby improving total manufacturing rate and procedure quality. It is time for both discrete manufacturers and OEMS to embrace the journey of Industry 3.0 along with Industry 4.0 through Industry 5.0 technological changes. Automotive manufacturing businesses give a purposeful and special enhancement of their operations by combining AR tools with advanced picture recognition modern technologies, computing power, IoT/IIoT tools, and AI to develop really powerful evaluating devices.

SUMMARY

Potential for industrial transformation in an automotive industry is substantial. Man, machine and manufacturing procedures are smartly networked individual products of top quality can be developed much more rapidly, and also expenses can be made competitive. The industry of the future will allow transformation at their origin to incorporate the consumer demand well in advance. It includes change of manufacturing processes including smart tools, reduction of manual labor, radical reduction in downtime and most significantly adaptable manufacturing systems. Because of the increasing demand for automation in factories, the future will certainly have more affordable cobots, software and small-sized workstations that are personalized. Therefore, not just the product but the devices making the product can likewise be modular, which would certainly customize itself based on the preferred product design change. Product design procedure in the future will certainly boost the present quality system to respond to consumers swiftly while maintaining the same level of quality and integrity. This will certainly not just include research into sophisticated materials but also include the processes that use them.

Industrial processes will certainly change as the future fads in the direction of electric automobiles, and also, independent automobiles will additionally impact production as well. The engine in the automobile is already being replaced by battery cells. This impacts the entire supply, and the process innovations in product design and development, product distribution and logistics also get influenced. Faster distributions come to be priority as customers strive for customizability and extremely customized products of the future. As machines become smarter with sensors, information collection and administration become an essential element of Industry 4.0.

Automobile industry is facing essential process transformation: the electrification of the powertrain, embracing the advancement of technology standards especially digitalization. Currently, most of the automotive manufacturing industries lie in between the second and third to fourth generation of industrial transformation. It indicates that although automotive manufacturers may have microprocessors, robotics and also computerized systems doing the support, they additionally have some type of manual work being done. It might entail people doing visual inspections on the product, product coordinators, logistics such as relocating containers manually; besides the low quality of components boils down to the individuals along with the absence of training in manual procedures. Therefore, automotive manufacturers need to go completely computerized in their journey from Industry 3.0 to Industry 4.0, before starting the drive toward Industry 5.0. Many aspects of the typical automotive sector are being influenced by the technological innovation; rather than anxiety, the SMEs plan to lead them by developing automobile parts that do not pollute, have zero emission, decrease waste and adhere to energy reliable processes that are crucial in this smart connected environment-friendly world. It unlocks brand-new technologies that revolutionize automotive sectors and often the culture itself.

BIBLIOGRAPHY

Akgun, A. E., J. C. Byrne, H. Keskin and G. S. Lynn. "Transactive Memory System in New Product Development Teams." *IEEE Transactions on Engineering Management* 53 (2006, Feb.).

Alkhoraif, Abdullah, Hamad Rashid and Patrick MacLaughlin. "Lean Implementation in Small and Medium Enterprises." *Literature Review* (2018): 100089. https://doi.org/10.1016/j.orp.2018.100089.

BMW UX. n.d. BMW Head Up Display: How It Works and What Information Can You See. https://www.bmwux.com/bmw-performance-technology/bmw-technology/bmw-head-up-display-explained/.

Ford. n.d. Building in the Automotive Sandbox. https://corporate.ford.com/articles/products/building-in-the-automotive-sandbox.html.

Henry Ford. n.d. 3D Printing & Product Design. https://www.henryford.com/innovations/education-design/3d-printing.

Hovorun, T. P., K. V. Berladir, V. I. Pererva, S. G. Rudenko and A. I. Martynov. "Modern materials for automotive industry." *Journal of Engineering Sciences* 4, no. 2 (2017): F8–F18.

Liu, L., H. Xu, J. Xiao, X. Wei, G. Zhang and C. Zhang. "Effect of heat treatment on structure and property evolutions of atmospheric plasma sprayed NiCrBSi coatings." *Surface and Coatings Technology* 325 (2017): 548–554.

Loughlin, S. "A holistic approach to overall equipment effectiveness (OEE)." *Computing and Control Engineering* 14, no. 6 (2003): 37–42.

Martin, J. N. 1996. *Systems Engineering Guidebook: A Process for Developing Systems and Products*. Vol. 10. Boca Raton, FL, CRC Press.

Mathivathanan, D., D. Kannan and A. N. Haq. "Sustainable supply chain management practices in Indian automotive industry: A multi-stakeholder view." *Resources, Conservation and Recycling* 128 (2018): 284–305.

Mayyas, A., A. Qattawi, M. Omar and D. Shan. "Design for sustainability in automotive industry: A comprehensive review." *Renewable and Sustainable Energy Reviews* 16, no. 4 (2012): 1845–1862.

Miller, W. S., L. Zhuang, J. Bottema, A_J Wittebrood, P. De Smet, A. Haszler and A. J. M. S. Vieregge. "Recent development in aluminium alloys for the automotive industry." *Materials Science and Engineering*: A 280, no. 1 (2000): 37–49.

Newsroom. 2018. Porsche Classic supplies classic parts from a 3D printer. https://newsroom.porsche.com/en/company/porsche-classic-3d-printer-spare-parts-sls-printer-production-cars-innovative-14816.html.

Piccinini, E., A. Hanelt, R. Gregory and L. Kolbe. "Transforming industrial business: the impact of digital transformation on automotive organizations." In *36th International Conference on Information Systems*. Fort Worth, TX, 2015.

Sturgeon, T. J., O. Memedovic, J. Van Biesebroeck and G. Gereffi. "Globalisation of the automotive industry: Main features and trends." *International Journal of Technological Learning, Innovation and Development* 2, no. 1–2 (2009): 7–24.

Ulrich, K. T. 2003. *Product Design and Development*. New York, Tata McGraw-Hill Education.

5 Transformation in Hi-Tech Electronics Industrial Sector

Electronics is one of the fastest developing, most innovative and the most affordable. Electronics sector plays an incredibly significant role in modernization, and a terrific emphasis on its advancement needs to be placed with the use of electronic technology in all markets of the dynamic industrial economy. It is composed of enterprises associated with the manufacture, design, development, assembly and maintenance of electronic equipment and components. Electronic items array from discrete parts such as integrated circuit, consumer electronic devices, industrial devices, medical and healthcare devices to information and telecommunication devices; besides, it supports a lot of manufacturing and industrial sectors. It has the responsibility to create all technically sophisticated digital devices for the future.

Electronic device capability together with material is expanding in vehicle infomercial and safety systems, manufacturing facility robotics and automation for industrial applications. Increasing technological innovation, consumer demand for smaller, much more effective tools with a never-ending amount of functionality over and above the fast proliferation of mobile devices are driving the development of electronic devices, market in a cost-effective manner. Significant changes can be seen in electronic product manufacturing that consists of a greater level of product combination, integrity, far better performance, increased number of products produced and minimized costs of device manufacturing. Original equipment manufacturers (OEMs) and original design manufacturers (ODMs) are significantly transforming product development process and new product development (NPD)/new product introduction (NPI) to electronic manufacturing service (EMS) providers.

Semiconductor manufacturers, from EMS providers to ODMs to OEMs and small to medium contract manufacturers, are frequently under pressure to introduce modern technology modifications to provide top-quality products. Manufacturers from the electronics industry deal with large challenges in regard to international competition with other suppliers. Extreme time pressure and diminishing shipment durations while being able to maintain constant quality at the same time suggest that optimal preparation of sources is necessary. Success in the hi-tech market typically depends on an organization's ability to supply innovative, affordable products to the marketplace, before the competition does. Digital suppliers have already dealt with this principle to change assembly line into completely automated connected factories. Industry 4.0 not only drives new modern technologies and smart products, but also additionally serves to broaden the manufacturing.

DOI: 10.1201/9781003190677-5

TRAIL TOWARD DIGITAL ECOSYSTEM

While steel was the main component of the initial industrial transformation, semi-conductors play a vital role until today and beyond in most of the industrial sectors. It is thought to be a crucial part of the next generation of electronic innovations; finding moral paths to sourcing products will certainly assist to drive the advancement of industrial digital transformation. Global consumers require more configure-to-order, make-to-order and assemble-to-order products. Mediocre processes and poorly integrated systems mean doom for EMS, particularly in today's age of rapidly altering consumer needs as well as modern technologies. Semiconductor manufacturers manage complex concepts to transform their own production lines right into completely automated smart connected factories. For the semiconductor market, the high cost of wafers makes attaching electronic components to the front opening universal pod totally practical and provides huge benefits to boost production performance.

The need for inexpensive, low-mix high-volume, high-mix low-volume adaptable semiconductor manufacturing EMS providers and OEMs requires leading-edge automation with control options that help consumers satisfy their business and technological objectives. Manufacturing of integrated circuits is very complicated and comprises numerous process steps, each influencing change to the silicon wafers at a microscopic degree. Semiconductor manufacturers consist of a variety of categories ranging from customer electronics to industrial to automobile. Traditional semiconductor production counts on a deal with the process recipe integrated with a timeless statistical process control (SPC) that is used to monitor the production procedure. Groundbreaking manufacturing procedures call for higher degrees of precision and accuracy, which require using tighter process control. Yield is attained by advance process control (APC). APC becomes an important element to boost efficiency, yield, throughput and versatility of the manufacturing procedure making use of run-to-run, wafer-to-wafer, within wafer and dynamic process control. Consumer short delivery and lead time demands are leading inspirations for adopting automation in manufacturing electronic facilities.

Automation in addition to integration is the secret to success in contemporary semiconductor manufacturing. Fabrication of semiconductor items demands innovative control on top quality, irregularity, return and dependability. It is vital to automate most of the semiconductor production procedures to ensure the correctness, efficiency of process series besides the equivalent specification settings, and to integrate all the fab tasks to provide effectiveness and integrity, together with the manufacturing schedule. Automation will give intelligence and control to drive the procedures of semiconductor manufacture, in which layers of products are deposited on substrates, doped with pollutants and formed using photolithography to create integrated circuits.

Electronic device production is increasingly affordable as significant investments in capital tools are called for to fulfill consumer demand for greater performing tools with higher functionality. Adding acumen to raw materials of electronic components promotes the totally decentralized operation designs connected with Industry 4.0. The advancement of standard electronics production is influenced by the gradual expansion of the global electronic devices market to emerging markets, along with the constant rise of process and labor costs related to electronics item manufacturing. Variables augmenting the growth of the automation market include the need for functional effectiveness, innovation, system assimilation and development in machine-to-machine (M2M) communication innovation. Electronic device manufacturers not only transform their very own facilities into smart factories to expand from design to manufacturing, but also develop new business versions as they launch their unique digital transformation.

PROCESS AUTOMATION REVOLUTIONS

Electronics component manufacturing and assembly is just one of the most intricate production environments in the industrial sectors globally. Most of the operations such as the fabrication procedures, assessment and product handling were done manually. As the dimensions of the elements reduced, device resistance to particle and other sorts of contamination became even more reduced, such that there is essentially no resistance to contamination on wafers, masks and others. Process modern technology has evolved to the point where low to zero levels of contamination have become the standard within the procedure and equipment settings throughout the EMS providers and OEMs.

Printed circuit boards (PCBs) are the core of any electronic- and microprocess-regulated powered gadgets over and above the most important required element of the electronic device industries. It creates the base for supporting wiring and surface mount of tiny components in electronics. PCB inhabited with digital components is called a printed circuit assembly (PCA) and the procedure of the assembly is called as printed circuit board assembly (PCBA). PCBA process is an extremely specialized and precision-based procedure to be carried out within the venture. The process has different phases including soldering paste to the board, picking and placing the parts, soldering, inspecting and screening. All these processes are called for and need to be monitores to ensure that the finest quality of the product is generated.

ELECTRONIC DESIGN AUTOMATION

The electronic industrial sector functions closely with the vertical sectors of modern technology such as automotive, aerospace, medical and other industrial markets that have detailed requirements. Digitalization of the hi-tech industrial sector has actually created a brand-new and expanding market for electronic design automation (EDA) for designers. Electronic component developers as well as service technicians were using a photoplotter to provide drawings of the circuit board, electronic components, and so on, which over time had actually been replaced by EDA. The flourishing automotive industry, Industrial Internet of Things (IIoT), artificial intelligence (AI) fields drive the development of the semiconductor market, which requires digital tools with complex design layouts. Automation makes provision for end users to enhance, customize and drive the capacities of digital design, test and verification using a scripting language and associated support energies. EDA helps in the specification, design, confirmation, application and examination of electronic systems. Such that, these can be produced either as an integrated circuit or multiple of them mounted on a PCB. In the automobile market, OEMs are buying EDA software application to establish the next generation of electrified, autonomous cars; in a similar way, in aerospace sectors, EDA capabilities are becoming more and more vital as avionic systems expand in intricacy.

> EDA presence in design is the expanding need from auto industry establishing a function called Advanced Driver Assistance System (ADAS). ADAS is driven by the development of AI, ML and Deep Learning (DL) advancements.
>
> *(Ansys, n.d.)*

Increases in the intricacy and electronic device content in the automotive and aviation industry are requiring changes in the computer-aided design/drafting (CAD) tools utilized to create electric distribution systems and wiring harnesses. Advanced silicon chips power the outstanding software application that has been used in day-to-day tasks in the business. They are the structure for every little thing from mobile phones and wearables to self-driving automobiles. Among the tough areas for the EDA market is radio-frequency (RF) design innovation. The RF integrated circuit (RFIC) designer wants to reduce prices; the goal is to obtain as many easy components onto the chip as feasible. Patterns developing in the electromechanical design of wiring harness systems include wire synthesis, auto-transmitting and automated generation of wire diagrams.

FAILURE PROCESS ANALYSIS

Challenges occur each day in digital design and manufacturing. Automation is the future of quality control. One crucial question that emerges is, as "Just how to review and maximize PCB design and manufacturing process." The simple answer is it can be accomplished by design failure mode and effect analysis (DFMEA) and process failure mode effects analysis (PFMEA). DFMEA is an approach for analyzing potential issues early in the product design and development cycle, where it is simpler to take action to get over potential problems, thus enhancing integrity with product design and substantially improving safety and security, quality, distribution besides expense. Fringe benefits from the DFMEA can be acquired through reason chain evaluation and mistake-proofing (i.e., poke yoke) to decrease risk priority numbers (RPNs). PFMEA helps manufacturing enterprises to build process safeguards to counter potential failures from occurring, i.e., determining the resources of failing and its influence on dependent variables via the manufacturing process. The RPNs prioritize the failure settings so that restorative activities are required to decrease the frequency and severity and enhance the detectability of the failure mode. Most of the manufacturers have actually been changing toward remote work, and the industrial economic situation in the years 2020 and 2021 is facing the appearance of new health risks due to the COVID-19 pandemic, which has forced organizations to rapidly adapt touchless manufacturing processes to satisfy compliance obligations to keep the labor force healthy and balanced. Therefore, quality management system or quality lifecycle management system enterprises can restrict the quantity of time as well as the sources they are taking into managing quality control.

Another important factor to consider while designing PCB by the EMS providers and ODMs is the decrease in product design cycle and the production expense, thereby enhancing yield. An important element of the product concept and design process is that an electronics-based item that can be effectively and cost-effectively generated is a design for excellence (DFx). It is an essential part of the NPD/NPI process. It helps in capturing everything before production and comes to be the liaison between the consumer and the product design team. Having DFx at a very early stage of design process, unneeded design and production holdups as a result of PCB manufacturer mistakes, tests access concerns, as well as out of date products, are got rid of, which ought to be the biggest value added for the EMS providers and ODMs, even for the small to medium contract suppliers. DFx targets the product value delivered to the customer, and it includes design for supply chain (DFSC), design for reliability (DFR), design for fabrication (DFF), design for assembly (DFA), design for manufacturability (DFM) and design for test (DFT). After the product design is done, the manner in which the PCB manufacturers perfectly incorporate the whole PCB production process and the use of PCB DFM evaluation helps to review and simplify the product design factors to be considered.

A glimpse on how PCB manufacturers can utilize automatic DFx tools to use their interior design rule checks (DRC) to supply comprehensive DFM records for design testimonials follows. DFSC is done early in the design cycle, which helps to determine the picked supplier component numbers' lifecycle state, accessibility, procedure compatibility and legitimacy that addressed prior to preliminary design. DFF assists to evaluate consumer designs as early as feasible, where it is very easy to make decisions that clear out price, improve manufacture returns and address issues before the final design is completed. DFA together with DFR reports plays a vital role to provide understanding right into product failures. Using Six Sigma, the estimated annualized price avoidances can be assigned to prioritize design modification decisions, and the evaluation can be executed to deal with issues early in the product design addressing prospective reliability problems. DFM integrates supplier detail equipment demands; particularly, manufacturing procedure demands are related to establish the process and eventually the control strategy to effectively deal with the DFMEA and PFMEA.

PLM assists to deal with leading DFR obstacles by automating processes, boosting exposure, access to vital information, and also engaging DFR early in the NPD/NPI process.

(Paganina and Borsatoa, 2017)

High quality product and product reliability is one of the most vital areas in the PCB manufacturing sector. It is vital to guarantee the quality of products, since the decision-makers throughout the product advancement decide on the high quality and price of the product. The details collected about design and process failures videotaped with DFMEA and PFMEA provide a really important understanding for future product and process design. Effective reliability engineering has the capacity to predict those parts of a product that may stop working along with the performance, safety, security, and financial effects of failing. Successful DFR is supported by effective product management methods. In the age of Industry 4.0, the arrival of IIoT along with product lifecycle management (PLM) systems produces a closed-loop, data-driven DFR program to improve predictability and integrity and attain better-performing products. PCB small to medium contract manufacturers, ODMs and EMS providers need to check out leveraging their product design and development making use of PLM, the system utilized to create their product, as a resource to connect other systems and data right into DFR to create an all-natural sight of the product development processes.

TRANSFORMATION IN PCBA PROCESS

Currently, the increasingly affordable electronic device sector, product demand drive increase the supply and and conversely boosts the rate, so the capability to integrate continually high criteria of quality and precision with optimized productivity, reduced manufacturing expenses and fast process times is important. The goal is to inevitably improve client worth as well as to preserve shareholder rate of interests with enhanced margins and better visibility on the market. Innovation has actually considerably enhanced the electronics manufacturing market. While incorporation and automation are transforming the means the electronics sector works, the industry has been slow-moving to catch up. EMS providers, ODMs and OEMs considerably gain from automation, which creates faster manufacturing, less mistakes and limited demands on humans. Limited tolerances and delicate nature of electronics made robot automation challenging, but development in robotic innovation enables manufacturers to understand the benefits of robotic automation and is among the largest segments of the robotics industry worldwide. Robots help in managing even the minute details with accuracy accuracy and substantially increasing throughput to reduce unit prices. The fast adoption of robotics helps in quick completion and enables EMS to be extra productive as compared to the competitors.

> Manufacturing of smart digital products, making use of microelectronics to shift from manual to semi-automated procedures fueled the adjustment from THT and also wave soldering modern technology to SMT.
>
> *(Whitmore and Ashmore, 2010)*

By removing delays, avoiding accidents and mistakes, improving management, as well as producing brand-new business paradigms, automation has significantly transformed electronics manufacturing industrial sectors. Through-hole technology (THT) assembly is used at first by hand soldering a wave solder equipment, wherein openings are drilled right into the PCB for installing of the components, besides solder is wicked up right into the holes to finish the circuit's connections. Advancement of automation in the electronics industries made manufacturing enterprises to begin utilizing surface mount technology (SMT), where digital components are put together with automatic equipment that places elements on the surface of a PCB. As opposed to traditional THT procedures, SMT elements are positioned directly on the surface of a PCB rather than being soldered to a cable lead. When it comes to PCBA, SMT is one of the most regularly used processes.

PLM FOR PCB DESIGN

Products are updated to new variations and capacities, and new components are included, lapsed and changed with totally new circuit layouts and innovation. Industrial markets across various domains acknowledge the vital duty that electronic devices play in the distribution of ingenious products, so effective management of the entire electronic component life cycle in the context of the end product is important. Managing product design, development, manufacturing and distribution are just one facet of taking a new item to market. PLM allows PCB designers to quickly take advantage of PCB data management performance while operating in the native electronic computer-aided design (ECAD) setting. PLM assists to find the ideal electronic device information promptly by making use of abundant data administration capacities and get rid of irregular and imprecise PCB component information by making use of enterprise-wide ECAD components' library administration. The practice of reusing parts has significant implications beyond merely reducing expenses related to design through the manufacturing job by recycling parts in NPD/NPI. PLM helps in automating and systematizing engineering procedures by integrating items and producing details, illustrations, thermal evaluation and simulation in the bill of material (BOM). Bi-directional data exchange in between PLM and PCB makes it possible for NPD/NPI teams to quickly connect with PLM in the context of their design layout environment, significantly boosting the capacity for engineering to leverage the PLM system to help drive product development ahead.

Business Challenge

Electronic engineers of small and medium contract suppliers and EMS needs to have accessibility to critical data such as life cycle, stock and rates during the initial design phase so they can make decisions in advance. Digital and electrical design layouts are based on detailed descriptions of several elements such as electric residential properties; supply status and information are generally held individually in CAD libraries, enterprise resource planning (ERP), manufacturing execution system (MES) and so on; however, there is a significant danger that design, sourcing, manufacturing specifications may get out of sync.

Precondition

The ECAD application is used for modeling PCB, supplier components and details about the customer's name, part number, sourcing information and ECAD library.

Approach

Product data management helps NPD/NPI teams along with PLM to successfully deliver tasks consisting of BOM monitoring that includes schematics and drawings, document monitoring creation and updating supplier data, configurable workflows and ability to track the job of the product advancement. With the technical innovation of smart connected products, PLM integrates with ECAD and mechanical computer aided design (MCAD) combines electronic devices data and design processes with mechanical data, so cross-functional teams (CFTs) can interact across design techniques and diverse enterprise applications. PCB design integration is the basis of electric design data management. By integrating ECAD with PLM system, product managers anticipate to minimize time to market, protect against mistakes and data storage problems and produce an extra skillful design review workflow resulting in better products. Having access to ECAD library helps to decrease product prices and help to conform to environmental regulations. CFTs share evaluation data in a digital environment throughout the extended group, therefore decreasing the need for physical models, shorten the development cycle and lower product development prices. A more vital process to consider is incorporating EDA with PLM, which helps in the reduction of product advancement time.

Result

The PLM provides security for the IP data while increasing design and development effectively by making it possible for PCB design teams to capture, handle, locate and reuse the ideal information from a solitary secure place. Tracks and takes care of product environmental conformity information within a single safe area consisting of IPC-1752 product material declaration throughout the product life cycle. Access to a full series of PLM capability makes it possible for the NPD/NPI group to take care of archived PCB data effectively and to optimize design through production processes. The electronics NPD/NPI team can maximize sources, reduce mistakes and project holdups and minimize total design costs. Regulated accessibility to all electronic device design information any time makes it possible for collective, multidomain codesign with complete traceability throughout the advanced product life cycle.

QUALITY ASSURANCE

The NPD of a digital product consideration is given to the functional life cycle, during which the product needs to function without fail, commonly in the form of a service warranty. Quality assurance is essential for all PCB as well as electronic device manufacturing. PCB manufacturing is in the eye of a speedy advancement today, much of it performed in service of miniaturization. Precise information that is legible besides being accurate is essential for board fabrication and assembly devices in PCBA manufacturing procedures, which can be helpful for traceability throughout the whole product development life cycle. PCB manufacturing is based upon the performance category as stipulated in IPC-6011 and IPC-A600 generic efficiency requirements for PCB. The use of digital twins and virtual testing permits product designers to obtain a complete picture of how the PCB, its parts and the end product are integrated and will operate in the real world. The quality lifecycle management helps EMS providers and ODMs to unify all top-quality-related activities across the supply chain for a natural understanding of product high quality and reliability. Quality management system (QMS) supplies automatic DFMEA and PFMEA and enables closed-loop corrective action and preventive action (CAPA) along with root-cause analysis (RCA) to accelerate recognition, control and analysis of concerns along with tracking of affected products. The QMS is needed to achieve conformity with regulative needs and high-quality standards. It integrates with PLM to become an eco-platform to retrieve and receive details and help identify problems early in the design process.

INDUSTRIAL ROBOTS

Robotics automation has terrific potential in PCB and electronics components making and applies to almost any type of phase of the entire manufacturing life cycle. PCBA requires extremely rapid, precise placement of small objects that are

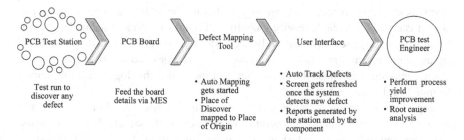

FIGURE 5.1   Automation defect mapping process in test station

often fragile. Industrial robots have the ability to carry out numerous jobs in turn, e.g., installing different kinds of elements on a base plate. It can manage display screens put together ports; build subassemblies; and apply adhesives, assessments, screening, packaging and more. Developments in grippers and vision technologies along with pressure sensors imply robots have to deal with a significantly large range of production, setting up and completing tasks. Force picking up allows for components to be finessed right into the location. When combined with flexible component feeders and vision systems, robots add flexibility to PCBA. Robotics assists PCB manufacturers with the adaptability to switch swiftly in between product variations.

For small and medium contract manufacturers, any type of gain in efficiency will certainly have a significant influence. As the cost of resources for chips, fiber cables, circuits and various other essential electronic device parts drops, manufacturers have actually turned to industrial robots to enhance operational efficiency and effectiveness and minimize labor costs without harming the quality and precision of finished equipment. Robotic assembly can adapt to the tolerance disparities, conveniently locating as well as adjusting the pieces as needed. With the decline in the setting up time, industrial robots are enhancing productivity in lots of electronics production facilities. Likewise, they assist in conserving cash on labor and manufacturing costs, as they pass those financial savings along to their customers. In PCB's too, there are new technologies happening in the robot sectors, in terms of size and much less programming. Tiny robots are utilized to construct automobile electronic control units, smartphone, PCBs, and so on, and to assist in testing and in the examination of tiny components.

Small to medium enterprise (SME) contract manufacturers are progressively looking to robots as a result of their ease of use and versatility, and their joint capability positions them well for automation. It is essential to understand that robots are not meant to replace workers but to make work easier for skilled specialists. As the industrial economic situation prefers a hybrid approach, safety and security is of primary concern.

PROCESS TRANSFORMATION REVOLUTIONS

Modern industrial innovation is helping various companies in various industries progress at a much faster rate. Hi-tech electronic device systems are the foundation for turbulent changes occurring in industries ranging from mobility to automotive to energy, to the fourth industrial change. Electronic industrial transformation has struck the telecommunication industry; the competitors in the field have made huge financial investments and advancement, leading to boom the industrial digital economy. Speed is crucial in today's service globally and thanks to modern technology, manufacturing enterprises have currently accelerated their service speed. 5G, autonomous vehicles, smart products, smart houses, smart cities and smart factories are in vogue, making it possible for electronics and modern technologies need to deliver unmatched degrees of reliability.

Ingenious electronic manufacturing is a main lineament of Industry 4.0, and also, manufacturing enterprises require to compete with one another by lowering expenses and enhancing performance by using innovation. The fact is that hi-tech electronics manufacturing welcomes a wider variety of activities beyond manufacturing; consequently, strengthening electronic production industries is important for sustaining international competition. The modern technical trend is moving toward smart and incredibly effective products, preferably with integrated safety features in addition to energy harvesting abilities. Complex products are manufactured based on customer demands by small to medium contract manufacturers, EMS providers, ODMs and PCBA businesses that have actually adjusted their operations to brand-new techniques of functioning. Similarly, integrated factories efficient in automation of complex as well as cost-effective products should make right investments in manufacturing facilities, transforming them into smart connected manufacturing facilities.

As technical innovations become quicker, transformations will inevitably follow one another in fast sequence in the coming years and beyond. The very first three industrial transformations took decades to play out whereas today's transformations last only as long as it considers industry-wide application to complete itself. Industry 5.0 integrates human employees, AI and manufacturing facility robots as they collaborate on designs and share work across PCBA manufacturing procedures. Developments in various innovations help EMS providers, ODMs and small to medium suppliers to move forward and embrace industrial transformation.

SIMULATION

Electronics and high-tech industries innovate at lightning speed to survive. Smart products have complex electronic systems that call for smooth real-world operations. Product developers encounter a challenge, as electronic devices are responsible for large scale thermal emissions; when a signal is sent out down a cable, it reverberates and discharges electromagnetic fields that interfere with various other parts of the product. Suppliers face a high level of variation because of, continuously shrinking batch dimensions and fluctuations in order to quantity that are increasingly difficult to forecast. With innumerable sensors, microprocessors and communication parts, product designers deal with tremendous product integrity and performance challenges with the miniaturization of gadgets, the support for multiple cordless technologies, faster information prices and longer battery life need to go through extensive need analysis. Simulation-led electronic improvement enables businesses to launch new products more quickly, at a lower cost with fewer resources. Engineering simulation plays an essential role in assisting the manufacture of cutting-edge and reliable products that accomplish and surpass target effectivity, energy efficiency, price and speed-to-market objectives.

Few simulations carried out are static, and dynamic stress evaluations can be executed for mechanical parts and casing structures. Liquid flow together with multicomponent thermal evaluations prevails simulation techniques for electronics parts such as chips, diodes, resistors and PCB, regulating the thermal emissions of various products, cooling effects and ecological impacts. Simulation in PCBA allows to determine production bottlenecks, highlights opportunities to raise throughput and determine financial saving opportunities such as optimization of straight and indirect labor. With the simulation of product performance in the initial design stage, NPD/NPI CFT groups will have the option to rapidly take in new modern technologies, with improved design and also much better materials, reducing the operation procedures and testing. It is necessary that all divisions involved in the electronic device manufacturing process should connect and also collaborate based on the exact same digital version.

> Few questions to be answered by EMS, ODMs, small to medium contract manufacturers before choosing appropriate simulation tool depending on the PCB functional requirement such as input signal, conversion of data from analog to digital, domain-based time and frequency sweeps.
>
> *(Peterson, 2020)*

Simulation technologies have improved and are integrated as part of the schematic capture program. It gives the PCB designer opportunities to check and also replicate the circuit essentially before proceeding to the PCB format and enables the NPD team to evaluate different materials for the elements and optimize layouts. Unlike traditional model testing, simulation enables engineers to practically examine just how a given product design will work well before any type of physical model is constructed against a wide variety of scenarios, some of which might be impossible to duplicate experimentally. Simulation spans the product design continuum to fuel open interaction between diverse design teams from electrical, mechatronics, mechanical to thermal and fluid dynamics. Simulation cannot build PCB; however, the outcomes will supply some useful insights into design modifications that can boost performance and fulfill customer requirements.

AUGMENT REALITY

Advancements in device equipment to smartphones are taking over the world like a tornado, and the hi-tech electronic industry has frequently proved itself to be at the lead of technology adoption. The development of advanced technologies such as AR and VR in the electronic industry is a significant revolution. These technologies simplify work for PCB designers in bringing the products to digital life, and they also make it quicker and safer for the manufacturing groups to assemble the elements of the PCB. Both the technologies are transforming businesses across industries and customer electronic device sectors, and the outcomes are effective. AR and VR address concerns such as fitting electronic packages into unusual forms and ensure circuit connections are working correctly while reducing the taxing procedure of the area and path in PCB production. In a nutshell, AR and VR provide PCB designers and manufacturers a true feeling of range and also closeness to have a far better understanding of the design early in the product life cycle.

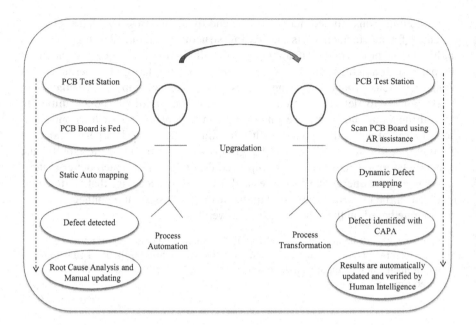

FIGURE 5.2   Defect mapping transformation using AR.

ADDITIVE MANUFACTURING

Additive manufacturing (AM) or 3D printing and electronic devices are highly connected. Personalization has become a huge asset while making use of AM for the creation of PCB and various other electronics items. PCBs are little; the procedure to model and manufacture them is rather prolonged. The arrival of 3D printing has actually introduced a new age for PCB development; it can create parts flawlessly adapted to a PCB for any type of electronic devices. It also clarifies complicated geometries that are difficult to make with various other conventional manufacturing techniques; moreover, it does not need any type of assembly procedure and also helps in reducing procurement expenses while removing any concerns about IP violation. A complex PCB can be created at a reasonably low cost, with the rapid turnaround allowed by AM; today, PCB board is quickly offered. It allows electronics component engineers to develop for functionality rather than manufacturability; those complex frameworks with ingrained electronics, enveloped sensors and antennas are readily manufactured.

Material choice is one of the essential factors to consider for an engineer when it comes to picking a PCB manufacturing method. Electrostatic discharge (ESD) is a real problem and a huge issue for the electronic industries, and having the ability to make ESD risk-free component is a real benefit to prevent any issue. ESD materials exhibit low electric resistance while using the required mechanical-, thermal- as well as chemical-resistant residential properties. ESD-safe 3D printing is used in jigs, fixtures and housings for electronic device making. AM is transforming every one of the methods by enabling suppliers to design and print jigs, fixtures and components with sophisticated engineering-grade materials that fulfill ESD surface area resistance demands.

AM printer from few manufacturers makes use of product jetting technology to publish multilayer PCBs with numerous features including interconnectors. Few business applications comprise of sensing unit modern technologies, radio frequency area systems and also IoT communication gadgets. Space sector has actually been a hefty adopter of the Polyetherketoneketone (PEKK) based ESD material in order to meet their chemical, warmth, and electrostatic discharge requirements for space trip.

(AFMG, 2019; Roboze, n.d.)

AM is positioned to play an integral part in production lines; extensive fostering has actually been obstructed by limited access to a wide collection of products that meet requirements for dependability, repeatability effectivity of 3D-published parts in the industrial ecosystem. 3D printing only utilizes products that are vital to construct the end item; much less material is used, which reduces the manufacturing costs, gets rid of waste and also minimizes the manufacturing time from a number of weeks to a couple of hours. 3D printing in the electronics industrial sector decreases warehousing and distribution expenses thanks to on-demand production and the possibility to create a digital stock. AM will certainly be a practical method for generating wearable, embedded sensors used in mobile phones and real-time health tracking. As the modern technology grows, expectations arise that 3D printing of electronics could ultimately shift from being only a prototyping device to direct end production.

Smart products made using materials such as composites, functionally rated materials, form memory materials, multiphase products, and biomaterials play a critical role in applications such as conductors, actuators, sensing units, soft robotics and wearable electronic devices. 4D printing innovation is in the pipeline, which will certainly facilitate the growth of electronic device manufacturing on plastic foils utilizing natural thin-film transistors, whereas boosted conducting polymers are being established for organic electronic devices.

ROBOTIC PROCESS AUTOMATION

Digital labor force is the backbone of most of the repeated processes these days. These digital employees include automated software bots, which are helpful for the back-office operations. Human reliance on Industry 4.0-based technologies brought to life the software application robots namely RPA. The most immediate effect of RPA is that regular jobs are carried out in an error-free, consistent fashion. There are a couple of areas where RPA assists PCB manufacturers It can be programmed to examine for broken traces over pierced and under drilled holes, misplaced parts and so on.

Among the key area in electronics industry is taking care of suppliers, where RPA review the invoices, extract data, using OCR, update supply details in enterprise systems (MRP) immediately, besides send notices to numerous needs for production planners to upgrade supply levels as needed.

(UI Path, n.d.)

The initial step toward implementing RPA is to identify extremely recurring tasks that are mostly prone to mistakes as well as think about piloting there. It plays a significantly essential function in small to medium contract manufacturers, ODMs and

EMS providers to transform into Industry 4.0. In this smart connected competitive world and difficult business landscape of high-volume, multistep processes with different authorization concepts and manual processes are automated from end-to-end with the help of RPA. Based on the business processes that need to be automated and their outcomes, enterprises need to select the process, followed by the readily available RPA tool in the marketplace, create, personalize and start implementing the option for automating the business jobs. RPA software program does not replace the already existing systems of the organization. In fact, they work in comprehensibility with the system. RPA is coordinated with any type of software application utilized by people, and also, it quite possibly may be carried out in a quick amount of time to accomplish functional strategies. Taking the following action toward Industry 5.0, electronics manufacturing enterprises need to take on a long-lasting process automation technique that aims to implement intelligent automation solutions that incorporate both RPA and AI capabilities.

Process Standardization of SMT

The business challenge is to obtain PCBA manufacturing functions to get in touch with the real-time information system making it possible for the anticipating technique in all features to provide greater stability and high-mix low-volume production environment.

Process standardization of SMT line results in

- Conveniently locate as well as keep an eye on key supply raw materials, final products, components and also containers to maximize logistics, preserve inventory degrees, stop quality issues and spot theft.
- Connect manufacturing facility assets and PLM, ERP, MRP, MES, DMT and MSD systems to provide role-based views using augmented reality experiences.

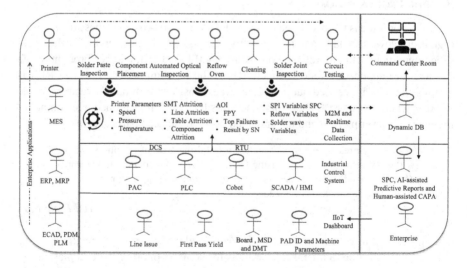

FIGURE 5.3 Process standardization of SMT line using IIoT.

- Enable real-time surveillance and predictive diagnostics of possessions to automatically trigger and proactively launch upkeep teams to reduce downtime and to identify maintenance and quality issues before they occur.
- Incorporate, assess and deliver insights from disparate and diverse silos of assets, drivers and also business systems right into unified real-time visibility of KPIs for enhanced operational effectivity and improved decision-making.

COBOTS

Cobots give electronics manufacturing enterprises the agility to automate nearly all the hand-operated tasks while adding worth to the businesses. Cobots make automation affordable and are a practical solution particularly for SME, EMS providers and ODMs as cobots are helping them compete better. Enabled by ML and geared up with innovative sensing modern technology, cobots work securely together with human beings, tackling dangerous, recurring and also significantly complex tasks. Cobots are redeployed over and over again in different duties to satisfy consumers raising the need for brand-new products, making them a valuable long-lasting investment and also a crucial innovation of the electronics industry. Semiconductor modern technology is making it possible for advancements in motor control, sensing and commercial interactions that allow cobots to function efficiently and securely on the factory floor. It coordinates with employees to highlight their finest and to transform the technological development in addition to a hike in top quality and productivity.

> Integrated sensors are completely suited for the delicate job of dealing with electronic components, securing delicate components and pricey fixtures, makes cobot a cost-effective, high performance automation device for PCB handling and in-circuit testing.
>
> *(Universal Robots, n.d.)*

The ML abilities make cobots to train various other cobots by sharing the details they discovered in-house as well as remotely from the cloud. With innovation transforming daily and manufacturing processes evolving perpetually, OEMs and EMS providers continuously have to adapt to the technical advancement of Industry 4.0 through Industry 5.0 to make manufacturing smarter, faster and more cost-effective products. Ideal security criteria are particularly crucial while implementing cobots; it comprises sensing units that allow it to be familiar with its surroundings, for fast, exact and safe operation. Data from these numerous sensors are refined rapidly, so the cobot reacts accordingly. With the arrival of AI, the cobot responds ever more suitably to the information accumulated from sensing units. This indicates that cobot can examine information, factor, resolve challenges and find out just how to respond to new scenarios, making decisions individually and interacting with the shop floor personnel.

Most of the electronic component suppliers are passionate about embracing the innovation since they will certainly operate in association with the workers in constrained rooms, production line without requiring any fencing and therefore conserving costs on the area. Cobots are an inexpensive modern technology with a much faster return of financial investment in the initial years of implementation. Small to medium contract suppliers, EMS providers and ODMs make the most of the possibility of using cobots with complicated applications in automation that increases the production quality while raising productivity with very little additional cost and also manpower requirements. Another essential element is designing cobots the electronic noise that interferes with electromagnetic fields, security and also ergonomics are some design challenges to be considered. The interesting thing is that as innovative technologies establish further, cobots will become ever better and more prevalent in the future industrial economy.

Artificial Intelligence

The artificial intelligence (AI) growth in the electronics industrial sector is fairly apparent. Powered by innovation as well as a flair for adjusting swiftly to arising trends, the electronics manufacturers have gone mainstream and fundamentally altered the method electronics components and end products are designed and developed. One of the most anticipated AI applications is using its potential in making the enterprise more anticipating and also flexible to a changing business atmosphere; this will assist electronic suppliers in creating a solid foundation for building cutting-edge electronic smart devices for the future. AI concentrates on bringing about significant modifications not just in the financial aspect but also in safety and actual operation control. It has become critical to include AI to get ahead of the global competitive market. The need to enhance client experience is high, and consumers are choosing tools that could provide even more personalized experience in terms of interaction and comfort.

The electronics industrial sector flourishes because of three significant advancement that comprises innovative analytics with insights, autonomous business procedures and AI-powered immersive experience that achieves more participation from the customer. AI computational power and advanced analytics at lower expenses can help small to medium contract manufacturers examine numerous information factors and historic information to expect machinery failure enables upkeep before it takes place. AI uses information to gather understandings besides spot patterns to determine sources of low yields and areas that require attention. Based upon the details, executing prompt and optimal modifications to production processes can enhance yields.

BUSINESS CHALLENGES OF PROCESS TRANSFORMATION

Technological innovation path is shifting toward smart and very efficient products that are with incorporated safety, security and protection features besides effective power harvesting capacities. Process transformation of EMSs and PCBA enterprises is vital to sustain competition and enhance manufacturing processes, reducing blunders and managing the manufacturing processes related to production and assembly of electronic end products. Electronics suppliers are urged to utilize their experience to establish not only their products, but also their own product design and process innovations. The link to the network is an important part of the process transformation. Speeding up the pace at which these procedure improvements are being made forces EMS providers, ODMs and also OEMS to constantly remain at the center of innovation in order to continue to be competitive.

Industry 4.0 through Industry 5.0 creates new trials and risks along with brand-new chances for electronics suppliers, who prepare to accept and invest in digital industrial economy. It will not only change an enterprise's own facilities into smart factories to expand manufacturing of new products but also build new organization designs as they embark upon their own business process transformation. Among the critical challenges for electronics firms is that it varies widely depending on the volumes, mix of items and running models: low volume, high mix; high volume, low mix; and medium volume, medium mix besides specializations, according to whom they sell to in industrial sectors such as automobile, energy, aerospace, defense or medical, and so on. Welcoming process transformation is also about embracing a mindset of frequently learning how to enhance production in addition to supply chain management and product distribution.

The effectiveness gained by the industrial transformation places electronics manufacturers in a possition where they have the ability to be much more active and versatile to respond to new possibilities. All electronic devices suppliers ought to be leveraging the data that currently stay across the whole supply to enhance specific processes in a fashion that contributes to the better the whole, instead of nearly adding innovations to their organization procedure. The success of smart process transformation is directly related to the acknowledgment of its added worth. Furthermore, it will be essential for firms to establish their basic strategic program at an early stage to gather experience matching innovations. It is expected that complete industrial transformation will certainly take electronic production services to an additional degree in regard to cost and quality.

Process transformation is likely to demand new abilities besides training that area greater focus on the communication between equipment and operators. Smart devices used in this transformation are successfully used to fix real-time issues. It is also most likely to ask services to believe in different ways about their use of industrial robots, cobots over and above how they handle their data. It assures boating of benefits for electronic devices manufacturers. There exist great deals of opportunities to improve operational effectiveness and boost performance by sharing workloads throughout electronic manufacturing procedures.

BRIDGING PROCESS AUTOMATION TO PROCESS TRANSFORMATION

Advanced process control and visualization applications are key parts of the automation to transformation strategy as well as inevitably have come to be the enabler for completely automated factories. The capacity for suppliers to track defects all the way to the PCB devices provides increased process presence through the celebration of digital data throughout the production line and also the channeling of this data straight to the manufacturing facility MES with venture applications. Advent of advanced process control and also visualization marks the very first steps toward a completely automated factory made it possible for by AI. PCB industry is advancing and also enhancing, not just in relation to product design through manufacturing and reliability, however, also in feedback to even more unified sustainable guidelines and commercial standards.

Sustainable Design through Manufacturing

Effective and reliable electronic component designs are contributing to greener electronics industrial sectors by decreasing energy intake in exerting to discover new and also much safer options to materials and processes that present a hazard to the environment and people. PCB suppliers are constantly facing difficulties in addressing waste generation; lowered performance along with increased power consumption always set back the techniques created to decrease prices. The methods used in electronic manufacturing industries are speeding up the transition of making in the direction of a sustainable system with production systems transformation and lasting value exchange.

> With ecological sustainability being a significant concern today, suppliers are turning their focus to just how they can use smart innovative technologies to come to be a lot more active as well responsive in terms of their environmental compliance, plans, and techniques. One of the methods remains in being effective with resources, lessening waste, looking at means to catch worth from waste, the greatest resource performance gains can be made.
>
> *(Esfandyari et al., 2015)*

Sustainable manufacturing is seen from different dimensions among the industrial sector which are environment, culture, economic situation, technology and performance monitoring. The aim of sustainability is to design and develop manufacturing process and component, wherein there is absolutely no impact on the atmosphere and accomplish 100% component recyclability. PCB manufacturers are currently concentrating on such merging to realize larger benefits of industrial transformation to attain sustainable production. Seamless assimilation of cutting-edge technology by Industry 4.0 that incorporated with Industry 5.0 creates huge amounts of information, which plays an essential role in establishing methods from ecological, societal and financial perspectives. Small and medium-sized contract suppliers mainly concentrate on energy effectiveness, performance, competitiveness, expense decrease and not on lasting manufacturing goals. They require to recognize the benefits of Industry 3.0 to Industry 4.0 via Industry 5.0 transform themselves right into a smart connected ecological community lined up with sustainable objectives.

SUMMARY

Developing a digital integration throughout PCB design via production is vital for generating end products that are premium quality, affordable and on time. Enterprises require to find out exactly how to make use of automation, AI system, ML, IIoT connectivity, data monitoring technologies to make electronic components more effectively, efficiently and openness. Electronics industry as a whole has been significantly accepting the IIoT. Spike is popular for electronic device products credited to the brand-new teleworking regimen throughout the COVID-19 pandemic, which has actually enabled workers of different industrial sectors to remain to satisfy the demands of working remotely. When it concerns supply and demand, electronics manufacturers require reliable and safe systems mostly cloud-based environment to keep operations, foster interaction within CFT as well as among vendors, representatives and retailers and take care of the inventories and item directories on an international range.

Based on functional requirements in the production process, the purchase division makes sure maximum degrees of supply, via the use of ERP systems for making process control work monitoring MES for PCBA process within the enterprise. Keeping huge data source of such elements consisting of non-technical information like minimal order quantity, preparations and pricing. Enterprises of all sizes will certainly be able to make educated choices based upon the real-time details that these smart connected tools can provide. Combined with other modern technologies of industrial transformation can aid drive a lot more effective short-term and long-term decisions.

Effectivity of the supply chain within the PCB manufacturers ends up being progressively based on vital modern technologies that sustain incorporated planning, logistics, smart purchase, warehousing and analytics. Sourcing of basic raw materials, supply components and final distribution of the item to the client, a shift to an electronic supply chain design uses the capacity for PCB manufacturers to take even far better control of their supply chain. Predictive upkeep stays clear of the prices associated with machine downtime, decreased maintenance and repair expense, which is improved by devices being more resilient. Coupled with data collection, industrial transformation will certainly help in forecast when and just how a piece of equipment may stop working, permitting enterprise decision makers to prevent it. As well motivated to use their experience to additionally create not just their items yet also their own design and process utilizing modern technologies, as the connection to the network is a fundamental part of the IIoT. Coming on to the interaction innovations wired or wireless, both will be an increasing number of sought after in the future and to have consisted of in the profile. Thinking about ecological destruction, policy waste electrical and electronic equipment (WEEE) needs to be implemented across the enterprise with the advent of technological advancement for the disposal of electronic waste in a secure way. Data and IT safety have an increasingly essential duty for the effective introduction of industrial transformation and need to be executed right into electronic systems as critical approval and success factors. Industry 4.0 and Industry 5.0 overall are still in a phase, in which it is essential to realize where and exactly how it can be applied with existing CFT strengths and innovations properly to fulfill the client demands.

BIBLIOGRAPHY

AFMG. 2019. All You Need to Know About Metal Binder Jetting. https://amfg.ai/2019/07/03/metal-binder-jetting-all-you-need-to-know/.

Ansys. n.d. Engineering Autonomous Vehicles with Simulation and AI. https://www.ansys.com/en-in/technology-trends/artificial-intelligence-machine-learning-deep-learning.

Bär, K., Z. N. L. Herbert-Hansen and W. Khalid. "Considering industry 4.0 aspects in the supply chain for an SME." *Production Engineering* 12, no. 6 (2018): 747–758.

Bassi, L. "Industry 4.0: Hope, hype or revolution?" In *2017 IEEE 3rd International Forum on Research and Technologies for Society and Industry (RTSI)*, pp. 1–6. IEEE, 2017.

Daim, T. U. and D. F. Kocaoglu. "How do engineering managers evaluate technologies for acquisition? A review of the electronics industry." *Engineering Management Journal* 20, no. 3 (2008): 44–52.

Daim, T. U., E. Garces and K. Waugh. "Exploring environmental awareness in the electronics manufacturing industry: A source for innovation." *International Journal of Business Innovation and Research* 3, no. 6 (2009): 670–689.

de Guerre, D. W., D. Séguin, A. Pace and N. Burke. "IDEA: A collaborative organizational design process integrating innovation, design, engagement, and action." *Systemic Practice and Action Research* 26, no. 3 (2013): 257–279.

Esfandyari, Alireza, Stefan Härter, Tallal Javied and Jörg Franke. "A Lean Based Overview on Sustainability of Printed Circuit Board Production Assembly." *Procedia CIRP* 26 (2015): 305–310. ISSN 2212-8271. https://doi.org/10.1016/j.procir.2014.07.059.

Krishnaswamy, K. N., M. H. Bala Subrahmanya and M. Mathirajan. "Process and outcomes of technological innovations in electronics industry SMEs of Bangalore: A case study approach." *Asian Journal of Technology Innovation* 18, no. 2 (2010): 143–167.

Paganina, Lucas and Milton Borsatoa. "A critical review of design for reliability - A bibliometric analysis and identification of research opportunities." *Procedia Manufacturing* 11 (2017): 1421–1428. https://doi.org/10.1016/j.promfg.2017.07.272.

Partanen, J. and H. Haapasalo. "Fast production for order fulfillment: Implementing mass customization in electronics industry." *International Journal of Production Economics* 90, no. 2 (2004): 213–222.

Peterson, Zachariah. 2020. PCB Functional Testing and The Role of Manufacturer Collaboration. https://resources.altium.com/p/pcb-functional-testing-and-role-manufacturer-collaboration.

Roboze. n.d. The Potential of Additive Manufacturing in the Space Sector. https://www.roboze.com/en/resources/roboze-additive-manufacturing-for-the-space-sector.html.

Shin, N., K. L. Kraemer and J. Dedrick. "R&D, value chain location and firm performance in the global electronics industry." *Industry and Innovation* 16, no. 3 (2009): 315–330.

Strange, R. and A. Zucchella. "Industry 4.0, global value chains and international business." *Multinational Business Review* 25, no. 3 (2017): 174–184.

UI Path. n.d. RPA Solutions for Accounts Payable. https://www.uipath.com/solutions/process/accounts-payable-automation.

Universal Robots. n.d. Melecs EWS GMBH Case Study. https://www.universal-robots.com/case-stories/melecs-ews/.

Wang, X. V. and L. Wang. "Digital twin-based WEEE recycling, recovery and remanufacturing in the background of Industry 4.0." *International Journal of Production Research* 57, no. 12 (2019): 3892–3902.

Whitmore, M. and C. Ashmore. "The development of new SMT printing techniques for mixed technology (heterogeneous) assembly." In *2010 34th IEEE/CPMT International Electronic Manufacturing Technology Symposium (IEMT)*, pp. 1–8, 2010. https://doi.org/10.1109/IEMT.2010.5746678.

Yin, Y., K. E. Stecke and D. Li. "The evolution of production systems from Industry 2.0 through Industry 4.0." *International Journal of Production Research* 56, no. 1–2 (2018): 848–861.

6 Transformation in Process and Industrial Manufacturing Sectors

Industrial manufacturing sector creates a selection of different types of equipment, from massive industrial to basic household products; some of the industrial products are product packaging materials, energy-effective unplasticized polyvinyl chloride (uPVC) products, glass, solar mounting panels, ground assistance devices, commercial valves, oil and gas and pharma products. Industrial manufacturing is a vast field to know. Efficiency is just one manner in which producing firms can distinguish themselves from others, and many ventures in this industrial section are conglomerates that create products located in numerous sectors. One major fad in this sector is using increasingly sophisticated manufacturing strategies. Small to medium range enterprises following distributed manufacturing development are expected to grow more as the global logistics market has become more complicated, and the manufacturers are better connected, as it has become a standard for the end product to be packaged and constructed close to the end customer. Gradually product transfromation leads in numerous businesses collecting data and operating in partnership with consumers.

Industrial manufacturing enterprises take advantage of making use of computer system-created settings that combine real life with digital threads. Current geopolitical conflicts and COVID-19 pandemic disturbances have actually reignited the argument on the future of globalization. In addition to distributed manufacturing design, warehousing along with logistics requires to be a lot more regionalized supply chains, including cloud computing and enhanced IT facility solutions to offer the supply chain the visibility needed to repond to market changes as well as proactively address the regulatory conformity issues that accompany international development to make use of expense and operational efficiencies. Industry 4.0 outfits the commercial production industry to strengthen its supply chains, to boost productivity, to improve procedures and to get a bigger share of the worldwide market.

SOLAR PHOTOVOLTAIC (PV) INDUSTRY

The renewable energy sector is the most needed sector at this time in the industrial manufacturing sector. Industries throughout the globe are trying to find means to generate more effectively without causing additional damages to the world environment. Future manufacturers, either small to medium enterprises (SMEs) or original equipment manufacturers (OEMs), are being educated on renewable power resources, encompassing the modern technology and the cost facets of renewable resource technologies, along with their potential which can give important insights. Industrial manufacturers

DOI: 10.1201/9781003190677-6

are huge customers of electrical power and a lot of them come to be a lot more lasting; they make great payments to promote a greener and cleaner planet. Knowing the regulations, certain power requirements, and the increasing relevance of power to drive the smart factory, there is more attention than ever before for energy effectiveness in industrial sectors. The fifth in addition to the fourth industrial revolution is about to bring even more changes in the future industrial economy.

Renewable energies developed an effective brand-new infrastructure for industrial sectors. Industry 3.0 had transformed the world a lot by producing renewable resource regimen, loaded by structures, partially saved in the form of hydrogen, dispersed using smart inter grids and attached to plug-in and zero discharge transport. Power consumption is a substantial contributor to international emissions besides climate adjustment. Industry 4.0 permits industrial manufacturers to change to renewable resources such as solar, wind and geothermal. Renewable energy plays a key function in the decarbonization of the whole universe. As renewable resource systems have actually become extra effective, costs have dropped substantially – a trend established with ideas such as smart metering and smart grids to proceed.

> Smart grids supply electrical power utilities, generators and users with the tools to connect as well utilize new technologies has actually resulted in a need for new power generation modern technologies and batteries, motivating stronger supply chains as well.

Consumers are requiring tidy, environmentally friendly approaches that cut down on carbon emissions to alleviate power besides energy expenses. Industry 4.0 makes it possible for have a reliable administration of an abundant yet unpredictable type of energy generation, supplying much-required stability and integrity. The favorable persons response to solar photovoltaic (PV) is making the industry more competitive. Solar PV is well on the path toward a levelized expense of electrical power with the correct application of smart innovation. Solar PV has actually been verified as a green power resource. The successful model of a monolithic, central power supply is increasingly being transformed to more flexible and decentralized. Energy collected from a variety of neighborhood sources is effectively coordinated; it is a lot more budget friendly, much more reliable and, certainly, greener.

Solar power systems such as solar ranches and concentrating solar power plants are becoming the globe's most important energy sources, generating more energy than other non-renewable fuel sources such as wind and hydroelectric systems, in addition to minimizing carbon emissions. Manufacturing industries make use of large buildings with a lot of roofing system areas as they are more suitable for a PV panel system. Resorting to solar will certainly save a lot on electricity prices while being shielded against the power price increase.

> Many industrial manufacturing enterprise leaders may assume that it is not inexpensive for small to medium-sized businesses, however, that is not real. As a company that utilizes a great deal of electrical power to power tools both exterior and interior lights, machines, and so on, the most effective way to manage the power prices is to locate alternate power sources, such as the solar power. The sun's abundant power is an endless resource of power that does not damage the ozone layer. Industrial solar power systems are an investment in the future of the planet that can help to utilize non-renewable energy resources and protect the environment. Solar PV production advancements

coming down the pipe are reducing the quantities of costly products such as silicon used in the manufacture of solar batteries, in addition to innovations such as bifacial components that allow panels to catch solar power from both sides.

New organization version Energy-as-a-Service (EaaS) is transforming the energy market. Businesses with sustainability objectives keen on extracting takes advantage of power financial savings, partner with an EaaS professional, who possesses a technology to assess the power profile of the business with the goal of determining the most effective opportunities for energy optimization. Power landscape is therefore transformed from being centralized, foreseeable, up and down incorporated and unidirectional, to being distributed, periodic, horizontally networked, as well as bidirectional.

Digital innovations such as artificial intelligence (AI), machine learning (ML), industrial robots, cobots, Internet of Things (IoT)/Industrial Internet of Things (IIoT) enable radical workplace transformations, maximizing human–machine communications and taking advantage of added-value human workers to the manufacturing operations. Technology progresses around the clock. Utilizing solar power in the automation industry is a technological change by itself, and industries are lucky to experience it. Solar energy is paving the way for ingenious and smart-connected modern technologies, particularly in industrial manufacturing. Solar PV capacity has broadened significantly as the rate of the innovation has actually dropped; however, the high expense of setting up stays an obstacle. Emerging SME's participating in manufacturing solar mounting structures turns into one of the biggest contributors to the decrease in installation costs.

PROCESS AUTOMATION AND TRANSFORMATION IN SOLAR ENERGY SECTOR

Increased use of renewables and resiliency problems along with issues of sustainability are just a few factors driving the solar energy sectors' demand to transform. Leading industrialized transformation economies Industry 4.0 and Industry 5.0 have a growing effect on the power field. Generation, distribution, consumption and smart manufacturing of solar energy are experiencing a massive revolutionary change as a result of technologies such as IoT, cloud and big data, ML, AI, augmented reality (AR)/ virtual reality (VR), digital twin, robotic process automation (RPA), industrial robots, cobots, blockchain, and so on. Transformation includes these technical advances to construct smart grids, to handle renewable energy, to distribute generation, to recognize usage pattern, dynamic monitoring and customer engagement.

The main question that occurs in our minds is this: why do solar power fields go for process transformation? Solar power markets rely on renewable energies; the disruption of COVID-19 has actually forced many to shut down their business, and it has underscored that the survival of many business depends on the implementation of automation. Decision makers need to recognize the optimal mix of services with a suitable, customized technology with those in market and an organization models along with system operations in an organized strategy.

Automation in Solar Power Plant

Solar power plant has hundreds of connected gadgets from a variety of suppliers distributed across the geographical area. Making use of a solar electrical PV system, it is necessary to evaluate just how much energy the system can generate according to location, positioning as well as plant conversion effectiveness. Incorporating some kind of performance monitoring system is important in order to keep an eye on the quantity of energy being created and to ensure that its responsds quickly when troubles take place. Utilizing a supervisory control and data acquisition (SCADA) system should be necessary to quickly keep an eye on, regulate and evaluate performanc to ensure that the projection conversion efficiency, low downtime and fault detection of a solar PV power plant will continue to be undamaged over its life span.

The SCADA procedure administration is exceptionally reliable with the visualization of real-time data with alarm systems. The interactive program framework for the simple representation and handling of area-sensing units, actuators and inverters is a beneficial element. Numerous screens and instinctive navigating on account of ordered views guarantee SMEs an organized introduction of the system operations. Advanced SCADA application analytics to the solar PV plant provides useful insight right into the plant performance. Providing these data visually utilizing graphs and reports on intuitive control panels helps to make the best use of efficiency as well as rundown details. SMEs can commence the automation journey with a versatile system by gathering and managing generation information from plant sites; SCADA systems sustain concerning every facet of a solar power plant.

3D-Printed Solar Panels

Creating new sorts of solar panels is a long process that contains various examinations and models. With enhancements in three-dimensional (3D)-printing PV panels, the modern technology is transforming much faster than ever. Using additive manufacturing (AM), cells can be manufactured in minutes permitting faster screening efficiency. Bolt mounting systems are few of the most commonly used products (hardware) in the solar sector, which is a terrific prospect for AM. In 3D-printed PV, the panel base is a transparent plastic sheet, set as layers from the semiconducting ink to the surface area, to create cells that are 200 microns thick, nearly four times the density of a human hair.

Solar power will ensure that solar manufacturing and shipment will fulfill the demand of the expanding green energy customers. 3D printing creates extremely thin solar batteries that can be published on economical products such as plastic, fiber or paper. Its ability to develop adaptable lightweight solar panels might have a higher favorable effect on future electronic devices, hi-tech apparel and even vehicle paint as well as paints used for structures in the form of solar spray.

RPA-Enhanced Solar Energy Sector

Solar energy industry is facing an age of challenges marked by intense competitors, stricter laws, environmental sustainability, and so on. Thinking about process automation as the primary step in the direction of Industry 4.0, RPA is one of the most fully grown segments of the broader arising AI market. Moreover, automation is now underway in the solar power market, beginning with management and repeated back-office procedures. Some of the locations where RPA are suitable are accounting, finance, human resources, audit and administrative assistance for procedures. With respect to SMEs, opting for RPA needs to start with creating a strategy, based upon the enterprise's existing IT strategy, and a high-level RPA process evaluation in order to confirm an extensive checklist of procedures and create the business situation to support financial investment.

Integrity, price and gaining back consumer trust drive the solar energy sector. Solar PV industries increase investments in RPA to improve operational performance through automating tasks, thereby helping to cut energy procedure expenses. The course to RPA ought to be a plan to grow business, incorporating AI with RPA as an important enabler. Both the innovations have the possibility to raise the precision of jobs, to reduce the labor strength of work, to carry out new analyses enabled by the ability to process, and to connect complex datasets. Decarbonization, deregulation and decentralization impact solar power market by allowing AI and RPA in managing the equilibrium in between demand vs supply, thereby improving efficiencies in all the entirety of the value chain, altering the consumer experience and changing service versions.

Incorporating AR in Solar Energy Sector

The AR lays over electronic details, straight or indirect sight of a physical, real-world setting; here, the elements are increased such as 3D versions and video clips, upon the real world through smart devices or AR glasses. Application of AR in the solar power sector has substantially benefited staff member safety along with property protection. Data analytics gives workable insights, and AR makes the data offered to the individual appropriate for making organization decisions.

Solar cell functions are common throughout; the capacity to imagine how systems look on industrial buildings before purchase is becoming vital for solar panel distributers. The operation and maintenance team, along with the engineering; procurement and construction team; and the asset management group are able to quicken existing operations as well as reduce the expenses too while boosting safety during the field operations. It assists to generate efficiency in staff member training, conduct faster upkeep activities and provide functional safety and security. AR, VR and mixed reality are removed in a huge method and bring in new worth to the energy sector. The enterprise is ripe for interruption from AR, which guarantees to make workflows a lot more reliable, effective, safe and productive. With AR systems having actually reached a budget-friendly price factor, these options that enable knowledge-sharing and make work environment productivity tools are excellent opportunities for SMEs to invest on the technological front.

Impact of IoT on Compact Linear Fresnel Reflector (CLFR)

Solar power plant is becoming a significantly practical alternative for energy manufacturing as typical power generation is becoming a whole lot costlier. Of the solar power plant innovation, CLFR is developing rapidly. CLFR is a sort of solar energy collection agency. It uses degree mirrors as opposed to allegorical mirrors that are used in solar parabolic troughs. The basic concept continues to be the very same with the mirrors collecting solar energy that creates vapor, which consequently drives a turbine. This modern technology leads to the manufacturing of vapor and not making use of warmth transfer liquid or any other medium. The sunlight that is concentrated with the aid of mirrors boils the water that is existing in the receiver tubes consequently creating heavy steam. No heat exchangers are used in this system.

Solar plants are metered in real time to determine their basic earnings; particular panels within a ranch are typically not checked. With the growth of the IoT, it is feasible to affix sensors to particular PV panels in a solar plant. The countless advantages contain granular real-time standing monitoring, real-time modification and likewise preparing for analytics. Overall, IoT will definitely improve the performance of solar plants in addition to making them much more available. Particularly, sensors will definitely enable solar energy plant supervisors to identify problems with information panels along with the layers of an IoT system.

SMART Solar Power Plant

If one is thoughtful about making use of solar energy for the industrial complex or the factory's power needs, two major points need to be considered prior to establishing solar energy: the solar meter connected to the solar energy system that checks how much electric power the system sends into a battery and the grid monitor that is a vital part of the solar power system that helps to monitor the battery levels by making use of a battery meter. Traditional solar power panels need to be monitored for ideal power result and lack the capability to determine; additionally, they take care of breakdowns in real time. With lasting energy systems instrumented, there are a variety of sensors used next to IoT-enabled networks, where a considerable amount of data is gathered besides being reviewed, consisting of PV panel temperature settings and extra opens up a brand-new standard to really feel, act, present and handle any element in a digitized environment.

As the digital interests of energy firms escalate, the need for improved connectivity to remote areas and the improvement of new ingenious Industry 4.0 technical usage is enhanced. By connecting the IoT with the solar energy system, energy providers handle each of the solar panel tools from one major control board that helps to recover reliable power end results from solar energy plant while looking for damaged PV panels, connections, dirt collected on panels and various other such problems that can impact solar efficiency. IoT helps link all the elements of power production in addition to consumption; it aids in gaining visibility at the same times supplies authentic control at every stage of power flow from usage to the supply throughout. Participation of IoT in the solar power system helps in monitoring solar plants as well as ensures ideal power result from another location dynamically. There is a lot more to solar than solar energy monitoring tools.

AI Application in Solar Energy Sector

AI and ML have the potential to evaluate the past, enhance the present and forecast the future. Boosted projecting and scheduling of power sources become important for the renewable energy industry in order to efficiently manage the grid. By integrating AI into the solar power system, sensing units affixed to the grids collect a big amount of data that provide useful information to industrial maintenance operators to provide higher control as well as adaptability to smartly readjust the supply with demand. Smart storage units can additionally be changed based upon the flow of supply. In addition to this, making a prediction about weather condition with the help of smart sensors and advanced sensing units will improve the overall assimilation and performance of renewable energy. In a nutshell, AI offers far better prediction capacities making it possible for improved demand forecasting and managing assets. Its automation capacity drives the functional quality leading to competitive advantage and increased financial savings for stakeholders.

> One of the industrial examples is sophisticated load control systems set up with the machine, such as industrial furnace, which can automatically turn off when the power supply is reduced. An additional instance autonomous drones with real-time AI supported evaluation will certainly come to be reliable and also effective analysis of photovoltaic panels.
>
> *(Gligor et al., 2018)*

AI applications can transform the renewable energy through boosted efficiency, which consequently will sustain the growth of the industry and ideally accelerates its fostering. Applying AI to the advancement of new products can decrease ingrained discharges, poisoning and costs. One of the most crucial variables to be taken into account with renewable energy is the reality that nature is unpredictable. Innovative technology is set to influence every part of a solar energy utility's operation. If used wisely, AI can become the most effective possessions paving the path to a cleaner and greener environment.

EXTRUSION PROCESS INDUSTRY

The introduction of smart cities, smart structures, transforming way of life, and so on, are a few factors that have actually given a rise to a new structure of modern technologies and products. The plastic extrusion procedure includes melting plastic products, requiring it into a die to form it into a constant profile and after that cutting it to size. Plastic extrusion is utilized to generate a large range of products across the industrial sectors, such as constructing products, commercial products, industrial components, armed forces, medical and pharmaceutical sectors. Pipelines, automotives, sports, window structures, electrical covers, fences, bordering and more are simply a few of the common things made by plastic extrusion. It has become a great selection for applications that need an end product with a continuous cross section. Plastic extrusion is one of the most extensively used production processes for the creation of plastics.

Every step in the extrusion procedure is essential for a specific quality of the end item. Product design groups need to begin considering extrusion needs while conceptualizing in the concept stage. Having an extrusion professional on the NPD team and participating in the brainstorming during ideation will enhance the conception process. The art of extrusion trusts experience and comprehensive quality control utilizing the right temperature level, speed, pressure, tension and time elements to develop a consistent product. Medical device manufacturers (MDMs) continue to make smaller, progressively intricate tools with unique geometries that need high-precision extruded components; effective shared interaction throughout the design phase helps the tool manufacturer to completely understand what is feasible and additionally offers the extruder a deeper understanding of the MDMs' assumptions and clarifying them with FDA's governing conformity.

Medical plastic extrusion is one among the effective techniques for changing the attributes of raw plastic utilizing a mix of ingredients. Medical procedures entail the transfer of liquids to or from the patient and employ a wide range of flexible tube products. Products used in medical plastic extrusion array from polyvinyl chloride (PVC) to polyurethanes to nylon copolymers to polycarbonate to polyether ether ketone (PEEK) to silicone. Its application includes catheters, syringes, dental tools, analysis instruments, drug shipment devices, implants, clinical bags and medical instruments. Extruded clinical items require the cautious application of precision handling ideas, specifically for microbore, coextruded or cross-head extruded tubes, for which the size resistances can be as tiny as ± 5 µm. Medical devices continue to need small-sized tubes with accuracy becoming increasingly important. The extruded medical tool design is categorized as tubes with solitary profiles, multilumen profiles, films for item packaging, sheets that can be post-formed into liquid containers, catheter tubing with encapsulated striping, multilayer tubes, films and sheets. Getting over new product design challenges needs to be a collaborative procedure between the MDMs and the extruders.

Raw material plays an important role in the plastic extrusion industry. uPVC also known as rigid PVC or vinyl siding or vinyl is among the most flexible and sustainable materials utilized in the construction industry. As it is totally free of Bisphenol A (BPA) means that uPVC can be used in medical as well as in dental equipment without the worry of contamination. State-of-the-art flexibility makes uPVC a perfect selection for windows and doors used in business, factory and domestic functions. uPVC doors and windows give reliable, effective thermal, audio insulation and help in energy preservation. Unlike timber and lightweight aluminum, uPVC maintains its shape in all weather conditions and stays unrestricted in case of any type of physical effect. uPVC is often utilized in dental retainers for its strong and non-toxic attributes. Growing popularity and demand for uPVC doors and windows have provided chances to numerous SMEs to endeavor into this industry.

Statistical Process control (SPC) is crucial in comprehending procedure capacities, recognizing unwanted variations, refining manufacturing procedures, and it allows enterprises to effectively and constantly fulfill customer's sophisticated demands for high quality, preparation, tolerances, distribution, as well as cost. Reviewing solitary process variables one by one is called as univariate evaluation, and it does not capture every one of the variables and communications influencing the high quality, whereas reviewing greater than one variable at a time is called multivariate methods. Multivariate data analysis (MVDA) has actually become essential for the continuous enhancement and upkeep of operational reliability. It is a statistical procedure for the evaluation of data including more than one sort of dimension. MVDA techniques are progressively being utilized for a range, and batch-to-batch contrast examinations to support and derive procedure understanding, which inevitably enhances the quality, security and effectiveness of medication components. With regard to the plastic supplier process, set points are marked within the procedure home window, and the robustness of the process can be checked by altering one process variable; depending on the initial settings, such examinations can bring about various process restrictions. Recognition of robust process setups is best completed with the design of experiment (DOE) strategies. DOE is an organized, efficient approach that at the same time investigates multiple process elements making use of a minimal number of experiments. Effective use of DOE can help with the advancement of models through making use of robust approach multivariate analyses.

The extrusion process is advancing together with the smart technological innovation. Sharing knowledge of how extrusion influences device manufacturing is critical to make intuitive choices that will ultimately reduce the time to market. The business model for industrial advancement offering innovative product designs, manufacturing and control solutions is Industry 4.0. One common variable across different industrial sectors is the transition to a green economy that will certainly have a global influence. SMEs' and OEMs' priority concern is about the dependence on plastic and its environmental effect. Plastic usage during COVID-19 pandemic has caused a resurgence. The process industries along with various other production sectors must strictly dedicate to environmental, social and corporate governance (ESG) plan focussing on waste generation management and reducing the damage on the ecosystem. Utilizing AI options with environmental management will certainly pave the way; the absence of strong emphasis and activity leads to the demand for a better technological option to conserve the atmosphere and increase sustainability with Industry 5.0.

PROCESS AUTOMATION POTENTIALS

A great deal of idea and process goes into making the uPVC profile, starting from the layout to the packaging. A uPVC profile is engineered to precision with all the life elements kept in mind. The fundamental part is the choice of the raw materials and the temperature they are being extruded. Three distinct stages associated with the uPVC manufacturing process are the formation of the material substance, extrusion of uPVC accounts and packaging and distribution. Applying Industry 4.0 to uPVC extrusion processes permits the assimilation of vendors and consumers along with the close interlinking of internal departments and procedures. Transformation of uPVC extrusion process to end up being an extremely specific fabrication procedure has actually continued thanks to Industry 4.0 and its allied technologies.

The uPVC extrusion is a continual processing approach that offers high-speed and high-volume production with the capability to develop profiles of differing forms, thickness and shades. Due to the complexity of the extrusion process, issues will eventually occur. It consists of raw material mixers, extruders, pass away, vacuum calibration systems, pullers, cooling systems and haul off. Each section of the profile extrusion procedure itself can develop one-of-a-kind troubles, leading to poor extrudate high-quality turns down. Some of the areas where process automation can be applied resulted in high effective operations, less wastage of raw materials, less rejections of profiles, energy efficiency, and so on as follows:

Real-time monitoring of information such as

- Raw material consumption
- Energy consumption
- Heating zone efficiency
- Temperature stability.

Thus, helping technicians take immediate action and allowing managers to monitor operational costs in real time.

- Construction of a digital business model of the production process
- Energy monitoring
- Process optimization
- Online quality monitoring.

PROCESS AUTOMATION TO TRANSFORMATION

Swiftly proceeding variables within the sector of plastic extrusion systems is the solution mode dealing with recycled plastic extrusion systems. Handling plastics clearly represents one of the most fragile stages in the worth chain, whether it is plastic molding, transforming, extruding or brightening surfaces. For the plastic molding industry, automation certainly does not result only in the automation of human workforce. The objectives of automation in plastic molding operations are to increase performance and savings. Technologies such as simulation, IoT/IIoT, data analytics, big data, AR and AM would certainly be relevant for process automation to transformation in plastic extrusion industries.

Updating the digital control systems permits better precision and uniformity throughout the high-quality production process with less wastage, and maintaining the temperature level consistently is vital to a successful plastic production process that creates top-quality items. The main parts of an extruder manufacturing process consist of a receptacle, barrel, screwdriver and electric motor. The second part is the raw polycarbonate material planned for extrusion. The last part needed for plastic extrusion is the die, which functions as the mold for the plastic extrusion. Heaters must be monitored, decreased, elevated or shutoff as required to maintain consistent warmth within the extruder, cooling fans and cast in heating system coats can additionally help preserve appropriate extrusion temperature levels. Enterprise decision makers worry about both their labor force and fulfilling the expectations of the customers; besides, they need to be mindful of the pressure that repetitive, non-value-added work puts on their employees.

Simulation in Optimizing Process Flow

Extrusion procedure is just one of the most essential manufacturing methods for creating ceramic, glass and polymeric products. Manufacturing has been performed based on empirical experience and trial-error techniques. Application of extrusion procedures is specifically prevalent in item manufacturing that makes use of polymers as the basic material. The temperature-level distribution inside the die alloy, in addition to the geometry of the flow channel, has considerable impact on the circulation behavior. Simulation is an effective means of checking out rheological defects, assessing, improving a procedure, positioned in the hands of the product and computational fluid dynamics (CFD) for engineers early in the product development life cycle as it allows very easy exploration of alternatives, delivering an improved end item, decreased scrap and tooling revamp costs. Simulation assists to determine dead spots, excessively long house times in the die and high pressure losses before the die is constructed. It enables optimization and design exploration to decrease waste and overdesign.

To anticipate behaviors such as fluid characteristics and mechanical stress, based on existing designs for the shaped parts, simulation helps SMEs to digitally mimicking virtually every element of the production procedure, from material circulation to coolant distribution to contraction and warpage of the molded part. Moreover, it offers guarantee for any plastic part developed for manufacture by stabilizing product efficiency in regards to toughness, rigidity and exhaustion life with minimum cost production. Development in simulation enhances and forecasts the development of the plastic melt front in addition to predicting and gauging the last shaped form with other specifications such as filling-up, analysis, warpage estimations, thermal optimization analyses and choices for customizing its product data source. Carrying out a mold flow simulation permits NPD groups to conserve time during the growth procedure, cutting prices significantly and attaining monetary benefits that are passed onto the organization. Computer-aided engineering (CAE)/CFD tools integrated with AI compute variants in the molding process beforehand.

AR Scope in Plastic Industry

Plastic industries are in the process of presenting themselves modern technologies to accomplish zero-defect production and improve work cell versatility. AR modern technology progressively expands its outreach by establishing its presence throughout various industrial fields. The key advantage of AR as opposses to other Industry 4.0 technologies is that it is very easy to test. AR is readily available and is budget-friendly for SMEs. AR projects are generally very easy to implement, and the systems on the marketplace use common plug-and-play methods that are reasonably simple to integrate into the enterprise's application ecosystem. One of the toughest challenges is knowing exactly how to integrate the pressing need for development with the ability of humans to adapt to something new. There are also lots of superior technological obstacles; the largest obstacle in bringing AR into the business will certainly be organization transformation. AR provides brand-new ways of interacting with manufacturing, product designs and machines.

AM Advances Plastic Manufacturing

The need for AM is enhancing day after day due to one-of-a-kind features as well as competitive advantages. Moreover, in the world of plastic molding manufacturing, 3D printing or AM plays a considerable function. One of the primary advantages is that it gives an eco-friendly plastic manufacturing approach that creates a product with minimal waste. Medical and dental items are regularly tailored to the specific using 3D printing. Machining the part from aluminum is a practical choice, and with 3D printing, one is able to effectively deliver polymer components in low amounts. Steel will not be the product of option, carbon fiber-filled 3D-printed polymer can replace even a tough or solid steel. Adopting commercial 3D printers to accelerate NPD by prototyping internal part is just a short leap away. With increasing manufacturing prices and the digitization of manufacturing, industrial manufacturing OEMs and SMEs continue to inconsistently evolve to keep functional dexterity, to keep costs down and consequently increasingly looking to 3D printing to remain dexterous, receptive and ingenious. While considering medical device segment, the geometric freedom guaranteed by AM and the capability to supply more personalized individual treatment cost-effectively is widely enticing. When paired with CT scanning, 3D printing is used to offer patient-specific services, such as implants and dental appliances.

Environment Management System

Carrying out an environment management system (Ems) in plastic industries will certainly aid to determine, analyze and take care of the ecological effects of the operations. Plastic is among the most favored materials worldwide. Plastics market ventures are confronted with a variety of challenges daily, and also, preserving the highest level of quality in their operational processes and products is a must. SMEs and OEMs of plastics industry swiftly see a return on investment when they discover and execute in conformity to the demands of ISO criteria for quality administration, security and environmental management. One such criterion in ISO 14001 focuses on the environmental effect that the activities.

An Ems focuses on waste minimization and reducing financial expenses from decreased waste, scrap, rework and energy usage. ISO 14001 does not state demands for ecological efficiency yet draws up a structure that an organization must comply with to establish an effective Ems. A solid commitment from the executive management level is important to ensure the effective implementation of an Ems. An initial review followed by gap analysis of the business process and products developed is advised, to help in determining all elements of the current operation and future procedures that might concern the environment. Plastics manufacturing industries that take a close look at their environmental effect inevitably discover opportunities to minimize waste by regularly generating substantial waste reusing procedure savings. In addition to enhancements in performance, SMEs and OEMs can gain a number of financial benefits including higher conformance with legislative and governing demands by adopting the ISO criterion.

Outcome of Transformation

Industrial transformation in the plastics industry is based not only on global connection but also on the arrangement of derived data along with process and digital improvement. One of the best challenges for automation industrial manufacturers is to outfit or upgrade all end consumers for the future and prepare themselves appropriately for prospective global industry needs. The idea is that the industry 4.0 sensation will certainly not stay a fad for the large gamers, but will become a living truth for small as well as for medium-sized enterprises. Having MES permits automated real-time precise data evaluation on consumers and materials from different vendors. MES is the maximized remedy for the SMEs plastics sector for cost-optimized and reliable link of the machines on a worldwide range.

Process optimization via automation is generally related to a reduction in scrap rates, downtime and better monitoring of manufacturers for predictive maintenance. Process transformation means a substantial renovation in the quality and the accuracy of data gathered and evaluated from machinery via the manufacturing processes. Unforeseen losses because of equipment disruption or wear down can be decreased with a variety of technologies consisting of sensors set up to gather real-time information incorporated with cloud that enables complex data analytics improved with ML. Operational tracking notifies anticipating maintenance; therefore, significantly decreases the downtime of the equipment as well as the prospective scrap from the process.

Abundant details derived from big data analytics provide insights as a result of the digitization of processes; this dynamically strengthens the understanding of the enterprise to take actions to achieve innovation improvements across the value chain. Every organization needs to carefully evaluate the advantages and challenges that Industry 5.0 through Industry 4.0 entail before starting the transformational journey. Business enterprise teams include environment, health and safety, quality and manufacturing. The senior administration, must continuously be included to improve the suitability, competence and performance of the Ems. Improving environmental performance will add considerable value to the organizations. With the introduction of Industry 5.0 SMEs, OEMs will be able to find waste creation as an additional opportunity and can transform waste disposal back into straight revenues drawing a path to a green and clean ecosystem of the future industrial economy.

GROUND SUPPORT EQUIPMENT IN AVIATION INDUSTRY

Airport terminals broaden their operations, resulting in the enhancement of brand-new terminals, entrances, traveler solution tools, along with another framework. Each gateway of an airport terminal calls for ground support equipment (GSE). The main function of the GSE is to support on-ground operations in between flights while the aircraft is parked at the gate to accommodate the needs of airplanes. The need for GSE is directly proportional to the growth of brand-new airport terminals and the addition of brand-new gateways, terminals at existing airport terminals. GSE is used to service commercial and military airplanes. Efficient, trustworthy airplane GSE is the key to guarantee smooth turnarounds and on-time separations thus avoiding delays, costs as well as trouble for guests. The growth of hybrid GSE, new batteries, quickly billing ports and improved electrical ground support tools with reduced maintenance costs is underway across the globe.

The main stakeholders in the GSE industry ecosystem are enterprises that supply lasting air travel gas, SMEs, modern technology providers, suppliers, providers, retailers and end consumers. The key ground power units (GPUs) of the GSE are battery-driven eGPUs, electricity-driven GPUs and diesel-driven GPUs. Considring the Ems, the battery GPUs and the electricity-driven GPUs are regarded as cleaner and greener whereas the diesel GPU leaves a large carbon dioxide impact. The Industry 4.0 technical era has the potential to boost GSE air transport key efficiency locations, where safety levels are so high in spite of the margins for enhancement being incredibly tight. The Industry 5.0 era could imply a change in safety improvement through maintenance, repair and overhaul (MRO). The successful delivery of airport terminal operations needed competent ground managing staff and GSE, which has dramatically boosted the need for aviation GSE around the world. Price has constantly been a variable that influences the selection of GSE manufacturers.

PROCESS AUTOMATION TO TRANSFORMATION

Rate, performance and precision are very important considerations in ground handling for decreasing the turnaround time. Efforts are taken to promote an eco-friendly GSE such as flight catering hi-lifts, ambulifts, self-propelled traveler step ladders, towable step ladders, luggage trolleys, aircraft test engine cells, cleaning trolleys, fuel browsers, containers and pallet dollys, mobile elevated observation systems, and so on, at airports. SMEs make use of Industry 4.0 technical development as the first step in transforming logistics procedures into event-based procedures looking for offered equipment dynamically to appropriate details such as fuel and temperature levels, so on to improve customer complete satisfaction besides gaining competitive advantages. Complete exposure in of GSE, for maximizing the ground handling solutions on the apron. By embedding smart sensing units on the GSE, SMEs will certainly be able to transform assets into important smart asset monitoring, thereby maximizing enterprise performance and processes. Time-consuming jobs such as GSE malfunctions help prevent blackouts and errors.

AR/VR Improve MRO in Aviation GSE

Aviation is a busy industry, and the important way leading airline companies, flight terminals and cargo business run smoothly is by leveraging the very best available in aviation GSE. As GSE suppliers and OEMs start to function closely with automated solutions, GSE training will certainly require to alter and provide the staff with brand-new understanding and abilities to guarantee safety. There are a few key challenges to be overcome to construct risk-free and efficient computerized GSE system. The pricey nature of the air travel industry amplifies the high cost of making errors. AR and VR are a terrific assets for GSE enterprises to provide a far better service and train the employees in a more precise means. It has become a game changer for aviation GSE technicians and MRO professionals to check different parts of the GSE. Industry 4.0 innovations are conserving money and time, boosting operational effectiveness and assisting to improve the level of service to their clients, thereby making procedures a lot more effective and improving the efficiency of MRO in GSE logistics.

MRO facility technicians assist to maintain the aircrafts and function securely and accurately by making use of aircraft ground support devices. Digital twins enjoy tangible advantages and have been extensively embraced in the GSE production process, which experiences modern technology is being fed into MRO. With the advent of Industry 5.0, GSE SMEs are relocating into more process-oriented worth than asset administration. Process improvement enables preventative and anticipating upkeep for GSE MROs, which increases integrity and safety. The dependability generated by digital twins can include dependability across the whole company value. An important channel for the digital twins is the sensing unit. All emerging assets all tend to have sensors, enabling ease of information that consequently provide insights together with the application of AI, which will certainly lay the foundation for a lot more accurate and thorough projecting. Digital twins along with AR and VR assist GSE SMEs to attain complete process transformation and automation. They help enterprises to do reliable predictive maintenance in the long-run paving course to reduction of costs, assets monitoring, decrease downtime as well as NPD. COVID-19 pandemic has actually created an urgent need for the promotion of industrial transformation initiatives.

AM Applied in Aviation GSE

3D printing's ability is to be adapted to generally anything that the filaments can readily develop. GSE will certainly benefit from AM technology; parts can be replaced quickly by printing them at any place by any person with a capable 3D printer. It is reasonable to theorize 3D printing as a sensible prototyping approach for a variety of elements for GSE models. Manufacturers of GSE can now own the market and gain access to legal rights to the CAD documents rather than keeping huge inventories of stock. Reduced weight, lengthy service life, peaceful operation and fantastic robustness are among the most important of those needs. It is possible that a substitute part could be all set within an hour of receiving the purchase order, rather than awaiting for gauged in days. It can be used to build full-blown mockups and is conveniently developed to exact specifications. AM stands to entirely change the GSE market.

OUTCOME OF TRANSFORMATION

Ground handling is critically crucial to the airline industry market yet it additionally needs to go for the lowest feasible cost. GSE ought to consider all safety requirements to be taken into account for the design of aviation ground support tools abiding by the ISO 6966 standards. Given the governing, ecological and open market conditions, it is easy to understand that GSE gravitates toward commercial improvement that provides a much better method to handle their daily operations, while also maximizing their resources. Smart asset tracking along with administration services aids to keep track of flight terminal assets and boost the effectiveness of ground handling procedures and maintenance regimens. GSE SMEs are realizing the benefits as all essential areas such as telematics data, safety and security, GSE vehicle parking, space usage, discharges and traffic researches can be better handled and understood with the arrangement of IoT information. Using electric-powered GPU or solar-powered battery-operated GPU is a fantastic way to go eco-friendly and remove carbon dioxide discharges. Appropriate improvement calls for numerous skill sets from radio preparation via standards growth to IT application with the best team assembled timescales and low cost. Industrial transformation in GSE sector is about getting rid of unneeded waste whenever feasible, getting rid of fuel usage, downtime and product waste are wonderful strategies to help accomplish bigger environment-friendly strategy outcomes.

VALVE INDUSTRY

Modern history of the valve sector parallels the industrial transformation, when Thomas Newcomen designed the initial commercial heavy steam engine that was subsequently advanced by James Watt, wherein vapor developed stress that needed to be included as well as managed, and valves got a brand-new importance. Valves plays a vital function in the high quality of our daily life, such as turning on a tap, making use of dishwasher, activating a gas pipe, stepping on the accelerator in the automobile. It is one of the most basic and crucial elements of our modern-day technical society and is necessary to all manufacturing industrial sectors and every energy production. Industries that rely on automated valves and tools include the food and beverage, OEM equipment, oil and gas sector, nuclear sector, petrochemical industry, shipbuilding, waste management and aerosols. Valves are a requirement for almost anything involving the activity of liquids and gases in a closed area. An additional important element of natural process is in the human heart with four valves to control the motion of blood via the ventricles, which keeps us alive.

Valves are straightly connected to the operational performance of a manufacturing procedure. It is vital for process designers and well as manufacturers to use a fresh eye to these elements as their evolving nature to complement the automation fad. The pipe system is not total without valves. Safety and service life span are the most important issues in a pipe process; it is crucial for valve manufacturers to provide premium valves. NPD approaches in valves industry have actually experienced lots of changes, but the fundamental design process continues to be unmodified. Industrial valve production procedure is a complex endeavor. Many factors contribute to its effectiveness: basic material procurement, machining, heat treatment, welding and setting up. Valves ought to go through extensive examinations to guarantee appropriate working before the producers hand them over to the end client. Modern market requires valves that have proven accuracy advantages and reduced labor and expense. Leveraging automation, which is simulation, procedure designers can figure out the minimum viable products (MVPs) of valves and examine these services making use of simulation to lower the time and sources, thereby bringing about the physical advancement of valves.

Fast improvements in innovation and the capabilities of computer-assisted control systems in addition to the integration of electronics have created smart automatic valves that has gained favor in the international market. Principles constructed around the IIoT have actually directed the industrial automation fields, fast fostering and mainstreaming of several manufacturing systems. The need for valves from healthcare and pharmaceutical sectors has increased during the COVID-19 pandemic episode. Process improvement and transformation is the most talked topic for management executives across the process industrial sector to help manufacturers automate and optimize their core manufacturing procedures and attain enhancements in other operational areas such as integrity, sustainability, safety and energy.

Process safety and security are the prime prospect for industrial transformation. A valve manufacturing enterprise will be able to keep track of the problem of thousands of control valves in a plant across the Internet by a connected solution based upon IIoT data collected. By utilizing data to determine very early indications, plants will be able to keep procedures closer to their optimum specification and arrive at better decision-making. Adoption as well as application of Industry 4.0 through Industry 5.0 required to get rid of a variety of functional and organizational obstacles, in many cases, because of the lack of modernization and automation in design and manufacturing.

PROCESS AUTOMATION TO TRANSFORMATION

As automation systems became more innovative, advancement in valve modern technology, performance and flexibility took a significant advance with the integration of electropneumatic control capabilities right into the valve. It improves valve manufacturing facility specifically address dimension and impact, power consumption, connectivity along with tracking of plant automation as it associates with upkeep preparation and functional effectivity.

Simulation Support Valve Manufacturing Industry

Valves are usually made for high-temperature liquids as well as gases that could affect their structural strength, because of too much tension generation and concentration in restrictive regions. Finite element method stress analysis is implemented to investigate the tension state of the valve body under numerous loading problems. CFD procedures recommends a variety of models for circulation speed, density, low-pressure areas around the bend, impingement angles for wear studies, minimum temperature level habits as well as chemical focus for any region where circulation happens. Product engineers design and model the performance of a whole system of pipes and valves to lower the possibility of failing. CFD simulation assists to check out the failure of aging infrastructure, providing designers a more exact picture of what had occurred. When valves manage high temperature of the fluids, the components certainly flaw under such high thermal stresses leading to the development of splits in the final end item and leading the valve to fail too soon. Transient thermal evaluation is done making use of CFD thermal simulation to forecast the early failure. Simulation assists NPD participants for a far better valve design optimization. With the development of AI and ML SMEs, component suppliers will be able to produce a generative design that can help in designing effective multiple variants of optimized valves.

IIoT in Valve Manufacturing Industry

One of the most important sections of process industry is the control valves. Valve manufacturers implement IIoT to boost and improve the performance of control valves and minimize upkeep price at every phase. The function of the control valves is to regulate procedure variables such as pressure, temperature levels and flow rates of liquid or gas. All these factors add to the general operational effectiveness of the process in the shop floor. IIoT links needed sensors that help in finding and managing different parameters of industrial valves in the field operation, which helps to appropriately keep an eye, regulate and handle the flow rate of fluids and gases in piping systems. Without a reliable control valve operation, the procedure would quickly become unmanageable for the operator. For conventional SME manufacturers, a much more vital challenge is that brand-new machinery cannot guarantee the creation of the exact same top-quality products that their customers have been utilizing for generations. IIoT offers condition monitoring, which helps SMEs to stop unintended downtime and improve valve efficiency. The evaluation of information obtained from IIoT-enabled smart valves help the enterprise's decision makers to make better decisions and boost the result.

Remote tracking of hands-on valves is another essential segment in process industrial sector. SMEs will certainly be able to achieve this by economical retrofitting investment making use of industry quality wireless sensing units and IIoT technologies. Predominantly manual operated industrial valves controlling large networks of process pipelines remain in massive use in chemical process sectors, paper industry and water drainage treatment plants. Generally, these kinds of sensing units are used for regulating the valves as commercial wireless valve placement sensors (angular placement sensing units and linear placement sensors) and detector sensors. The sensor gadget then reports the placement information in a digital layout to the SCADA main control system with field instruments, which will certainly be kept an eye on by dynamically using an IIoT system. The transformation of procedure control valves takes long period of time to roll out, due to high retrofitting investments. The enterprise enjoys the transformation benefits converted into cost savings, far better security and continual procedure optimization.

IIoT allow industrial valve production users to gather and also keep information from mostly all assets at an extremely high frequency and also at exceptionally low cost, which enables effectiveness in work besides business procedures additional to DCS and PLC. SMEs, component manufacturer and OEMs able to monitor the levels, temperatures, usage, waste, OEE and predictive upkeep data making use of IIoT. The IIoT platform gives boosted visualizations which allow operators to see modifications in the system together with atmospheric change much faster than usual.

(Reynolds, n.d.)

Third industrial revolution brought innovations in hydroelectric power industry and began incorporating information and communication innovations in its power plants and power grids, where automation along with digital governing controls began to form. With the onset of Industry 4.0 and Industry 5.0 across the globe, there is an enhancing requirement for eco-friendly plants aligning with IEC 61850 power plant automation control standard and an increasing variety of energy sources, both renewable and non-renewable. Of which hydroelectric power is an attractive type of energy because of its reduced carbon emission, affordability and, of course, the wealth of water. As with any energy generating procedure, control and surveillance software is a vital way to keep plants controlled. DCS permits the facility to constantly recover and analyze plant efficiency information, checks key performance signs, provides workable details for plant employees and provides necessary information when required in real time. Measurements of temperature, pressure, resonance and other parameters occur at localized sensing units, which are transformed into time waveform signals and examined by the plant operator. Power plant engineers have the alternative to keep track of the overall system condition and assign sources to deal with concerns as they are predicted and prior to part failing. The IIoT platform got in touch with SCADA, and PLC assists to readjust device parameters to maximize the results. By implementing IIoT, the hydroelectric industries will maximize their upkeep procedures, cut costs and strategies.

SUMMARY

Process transformation in the process and industrial manufacturing industry through real-time monitoring and optimization of control valves, lead to efficient upkeep processes, elimination of unstable manual treatment, far better employee safety and security and reduced production costs. Industrial transformation is changing the method organizational leaders assume and how they collect and utilize information to optimize procedures. Automation innovation is much more sophisticated than anything that has preceded; robotics and automation systems are coming to be antiquated and can present safety and security concerns. AI and robotics, in cooperation with brand-new innovation like 3D printing, show advantages in bringing advancement in production and satisfying raising consumer demands. Data analytics make possible for suppliers to pivot from preventive to anticipating maintenance. The mix of traditional process control systems and new technological innovation is the foundation to improve the accessibility of details besides boosting decision-making. To stay competitive, enterprises need a system that will certainly manage the needs of clients, suppliers, management executives and other stakeholders in a seamless way, which will certainly be enabled by Industry 4.0 via Industry 5.0 that is readied to take root across the process and industrial manufacturing community. SMEs, OEMs and part manufacturers take the effort to recognize and harness the power of process automation as well as process transformation to stay on the zenith of the new digital era. Industrial transformation is a great journey where the processes and industrial manufacturing sector advancements become ingenious, flexible, data-driven and step into the future eco-driven industrial economy.

BIBLIOGRAPHY

Ang, J. H., C. Goh, A. A. F. Saldivar and Y. Li. "Energy-efficient through-life smart design, manufacturing and operation of ships in an industry 4.0 environment." *Energies* 10, no. 5 (2017): 610.

Calik, K. and C. Firat. "Optical performance investigation of a CLFR for the purpose of utilizing solar energy in Turkey." *International Journal of Energy Applications and Technologies* 3, no. 2 (2016): 21–26.

Faheem, M., S. B. H. Shah, R. A. Butt, B. Raza, M. Anwar, M. W. Ashraf, M. A. Ngadi and V. C. Gungor. "Smart grid communication and information technologies in the perspective of Industry 4.0: Opportunities and challenges." *Computer Science Review* 30 (2018): 1–30.

Gligor, Adrian, Cristian-Dragos Dumitru and Horatiu-Stefan Grif. "Artificial intelligence solution for managing a photovoltaic energy production unit." *Procedia Manufacturing* 22 (2018): 626–633. ISSN 2351-9789. https://doi.org/10.1016/j.promfg.2018.03.091.

Gorecky, D., M. Schmitt, M. Loskyll and D. Zühlke. "Human-machine-interaction in the industry 4.0 era." In *2014 12th IEEE International Conference on Industrial Informatics (INDIN)*, pp. 289–294. IEEE, 2014.

Huang, Z., H. Yu, Z. Peng and Y. Feng. "Planning community energy system in the industry 4.0 era: Achievements, challenges and a potential solution." *Renewable and Sustainable Energy Reviews* 78 (2017): 710–721.

Kulichenko, N. and J. Wirth. *Concentrating Solar Power in Developing Countries: Regulatory and Financial Incentives for Scaling Up*. Washington, DC, The World Bank, 2012.

Lin, K. C., J. Z. Shyu and K. Ding. "A cross-strait comparison of innovation policy under industry 4.0 and sustainability development transition." *Sustainability* 9, no. 5 (2017): 786.

Pozdnyakova, U. A., V. V. Golikov, I. A. Peters and I. A. Morozova. "Genesis of the revolutionary transition to industry 4.0 in the 21st century and overview of previous industrial revolutions." In *Industry 4.0: Industrial Revolution of the 21st Century*, pp. 11–19. Springer, Cham, 2019.

Reynolds, Peter. n.d. IIoT Enables Control Valve Maintenance Improvement. https://www.arc-web.com/industry-best-practices/iiot-enables-control-valve-maintenance-improvement.

Rosin, F., P. Forget, S. Lamouri and R. Pellerin. "Impacts of Industry 4.0 technologies on Lean principles." *International Journal of Production Research* 58, no. 6 (2020): 1644–1661.

Schütze, A., N. Helwig and T. Schneider. "Sensors 4.0–smart sensors and measurement technology enable Industry 4.0." *Journal of Sensors and Sensor Systems* 7, no. 1 (2018): 359–371.

Stock, T. and G. Seliger. "Opportunities of sustainable manufacturing in industry 4.0." *Procedia Cirp* 40 (2016): 536–541.

7 Upgradation of Industry 4.0 to Industry 5.0

Industrial transformation and its technical advancement offer us a brand-new paradigm in manufacturing across different industrial sectors. Industry 4.0 has gained ground; many manufacturing enterprises comply with the path in the direction of process automation by means of digital improvement throughout the extended manufacturing processes. Customer expectations, the advent of smart connected machines and systems are driving the continuous digitization of production. Industry 4.0 has allowed manufacturers to increase operational visibility, reduce costs, quicken manufacturing times and deliver phenomenal client assistance. The onset of the COVID-19 pandemic has made business tycoons continue to embrace modification in order to keep ahead of competitors and win market share in an ever-developing industrial transformation. Manufacturing enterprises that measure their agility and automation degrees, for instance, could discover that given their degrees of manual work, these are the remedies they should concentrate on now, whereas their time-to-market modern technologies still serve them fairly and nicely. Industrial transition across various industrial sectors is complicated, regularly advancing at a breakneck rate and will certainly present the enterprise with new challenges that start with defining the major service requirements, the difficulties to fulfilling those demands, the potential services to fix those challenges and that need to evolve the industry from current to the future state. As the market goes on, the value chain develops with it and continues to remain in steps with updates and brand-new forecasts.

BUSINESS NEED FOR INDUSTRIAL TRANSITION

Modern industry has actually seen significant advancements since the start of the industrial revolution. To compete internationally, it is essential to embrace the evolution of technology and recognize exactly how to harness its power to enhance business operations. To start, it is essential to comprehend along with details the direction they are heading to. Manufacturing enterprises can build on their existing methods and technologies can make them more powerful with the help of Industry 4.0 innovations. Making use of these opportunities calls for substantial financial investment. A clear business strategy will certainly be essential to obtaining support from stakeholders, thereby safeguard the required spending plan. Initiating an organization transformation approach enhances the possibilities of prospering in this new digital age. Manufacturers ought to not consider digital transformation as an expenditure, but instead a revenue; enjoying long-term objective will improve the operational effectiveness along with productivity.

In addition to organization transformation, fully committed management executives who are accountable for the digital transformation and the integration of

DOI: 10.1201/9781003190677-7

brand-new items, systems and solutions are important. Without the ideal management, the reform initiatives might be obstructed. Technical innovation plays a tremendous role in obtaining a competitive advantage in a moderated industrial sector and thus must be recognized to be a force that drives industry competitors. Customers nowadays can shop around and often compare the local products with their global counterparts; this affects the SMEs who find themselves among international competitors even if they do not import or export items or solutions. This has of course a massive influence on an enterprise's approach in order to stay ahead of the international competitors. To access the impact of globalization, the manufacturing enterprise has to comprehend that there is absolutely nothing like one consistent means to determine it. Instead, several approaches in the direction of gauging globalization have actually been developed over time.

Considering the degrees of inventiveness, it is also essential to recognize what the driver of inventiveness generally is. As the product lifecycle reduces, companies have to act when necessary. Technology is a driving force for generating business value rather than simply playing the sustaining role. Consequently, enterprises had to enhance their rate of innovation, which has led them to pursue new services and product developments faster than ever before. For manufacturers, the concept of the smart manufacturing facility of the future is becoming true; besides, those hesitant to accept these advancements are discovering it hard to disregard. To obtain an understanding on how to boost the process together with operation standardization and comprehend where to purchase terms of technology, the business enterprise needs to recognize their weak points along with their business process factors.

The manufacturing industry's decision contemplate the future industrial change. Industries across the globe are transforming quickly, the enterprise's decision-makers are driven by the need to remain ahead of their competitors and face two essential questions: Do we have an option to embrace or not embrace the transformation? How long can we wait before transforming? Actual transformation is only just when the typical means are tested and the new path is complied with. Operating in repetitive processes, beginning small and scaling up is crucial for success. Considering the COVID-19 pandemic, it is an ideal time to pursue the industrial transformation journey from Industry 3.0 to 4.0 to 5.0, taking into consideration safety, environment, health and wellness and revenue.

CHALLENGES IN INDUSTRIAL TRANSITION

A common adage says with great opportunity, there comes great challenges. Industrial shift continues to transform the methods of SMEs and original equipment manufacturer (OEM) across the world, and brand-new challenges occur. For manufacturers, efficiency is vital and the point of discomfort comes in several forms. Most of the manufacturers rely upon breakthroughs in automation on computers and electronic device-aided innovations; information and communication technology (ICT) assist to gather, assess and give helpful information in the real time to the production systems. Industrial transformation is changing the way goods are intended, designed, manufactured, serviced and made environment friendly. Applying advanced modern technology in an existing production system, the economic evaluation and return of investment (ROI) along with the return of value (ROV) requires to be evaluated extremely carefully. The threats connected must be computed and taken seriously. Workers need to acquire brand-new collection of skills to fill the void to pursue improvement. Pressing research and development in such fields are also important. So, the additional investment that needs to be made to take on the more recent modern technology would certainly be compared with the losses in production during an upgrade along with the time obtained to recover the ROI with the earnings within existing system that affects the adaption of newer modern technology.

The main challenges are the unstable market demands, the need for better and faster manufacturing procedures, margins that continuously reduce and intense competitions in between companies that no organization can win without the help of smart connected innovations. Many sophisticated manufacturers are currently leaning on multiple traditional data systems such as enterprise resource planning (ERP), material requirements planning (MRP), manufacturing execution system (MES), product lifecycle management (PLM), supervisory control and data acquisition (SCADA), programmable logic controller (PLC) and Robots. Solutions that allow the seamless integration of the data systems are the trick to attaining success. The crucial concern that develops is the ideal means to measure key performance indicator (KPI) or the success matrix of the industrial transformation. A massive number of data are generated from different machines in the shop floor, so recognizing how to leverage particular data for the enterprise or certain KPIs are vital when on identifying the ideal technological tools that suit the business goals itself difficult. To conclude, manufacturers need to produce an out-of-box retrieving strategy to manage failing situations. The ever-changing fast pace of industrial cutting-edge technologies will certainly encounter lots of brand-new obstacles and will certainly continue to evolve in time.

MITIGATION OF INDUSTRIAL TRANSITION CHALLENGES

Manufacturing enterprises that emerged as success stories in the third industrial transformation are the ones that strongly welcomed change, faced the threats, recognized as well as properly embraced the favorable innovations triggered by the pertinent drivers and developments seen around them. It could differ from OEMs, yet can be a vision for SMEs. As a mantra begin of with examining enterprise existing state of maturity followed by identification locations where to boost maturation efforts, which lays the structure for successful initial action in the direction of change. For SMEs, Industry 3.0 values is a great place to begin, starting small helps limit anxiety, promote adoption and secure management executives buy-in which is another landmark in this stunning transformation journey. Few SMEs lack resources to assess the technical maturity of the pertinent technology as well as their service uses. Some SME management executives lack a systematic method of application. In order to fill the ability gap, the SMEs need to, develop and discover programs that support and integrate new concepts with hands-on possibilities that transform their manufacturing procedures and domains. Keeping an eye out for the smooth assimilation of the traditional applications with the upgraded modern-day technology is necessary, so that everything rolls under one solitary structure without affecting the major business process. Integration can take care of automating existing manual tasks and processes.

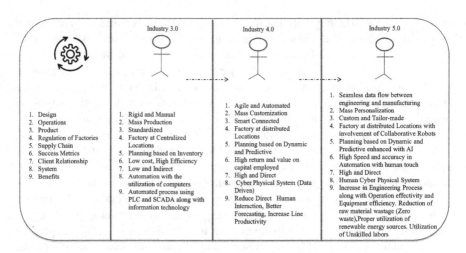

FIGURE 7.1 Transformation journey based on different factors.

The first expedition phase is necessary and when carried out with the ideal focus can bring about clear and concise final thoughts. Information administration clears up possession, and privacy. Data exchange between businesses makes it possible for third parties to acquire an understanding of the organizational techniques; so, it is to be cleared as to whom the created information belongs to as well as who is qualified and exactly how to use them. Take a look at the options to executed and adhered to the team capacities of the internal group or keep an eye out for the vendors' capacities to service the requirements. Create a durable system architecture prior to the execution that stabilizes information technology (IT) demands and the storage area whether cloud or in-house storage space. Production line operators need to be more involved, along with cross-functional team (CFT) of new product development (NPD)/new product introduction (NPI), the ICT team, and the management executives have all the essential details for higher responsiveness, liability and possession, thus making complete engineering to manufacturing supply chain presence.

Manufacturers practice to take into consideration business value innovation and new revenue streams as they utilize the deluge of data produced from innovative modern technologies. Cognitive devices and sensors are integrated to the production line physical systems by incorporating the capabilities of Industrial Internet of Things (IIoT), big data, 3D printing, analytics augmented reality (AR), virtual reality (VR), artificial intelligence (AI), machine learning (ML), cobots and human knowledge serve business functions. Human involvement is typically needed in previous, existing as well as future industrial transitions. Communication between human intelligence with computerized systems is anticipated to take manufacturing to new levels of optimization and automation. New abilities will certainly be called for in the locations of smart systems, robotics programs, new arising innovations and creativity. The essence of Industry 5.0 is applying innovations in the industrial sectors to accelerate their efficiency and not to replace the humans. The application of industrial transformation in modern technologies guarantees effectiveness, and optimal performances will be attained with minimal impact, influencing both the leading and bottom line of organizations.

INDUSTRIAL TRANSITION STRATEGY PLANNING

Industrial transition is so disruptive to standard business models and traditional concepts of industry competitors. Assess the situation of a business enterprise by looking into its structure that assists industrialist and entrepreneurs shape their approach toward success. There exists many helpful strategies management planning tools for SMEs to stay one step ahead of their competitors in this smart global industrial sector. Modern cutting-edge innovations are currently the factors driving business purposes as opposed to just playing a supporting duty. Industry evaluation is a critical component and is developed to assist enterprises to determine how it operates within an industry. SMEs need to recognize their standing among competitors through detailed market forces on a worldwide level and to analyze the standing and potential profitability. It is much better for an enterprise to have a strategy starting with solid gap analysis to deep dive into the transition.

The secret of establishing an affordable business strategy is to recognize the resources of the business, identify its crucial strengths and weaknesses and transform areas where strategic changes will certainly cause the best rewards. In the light of the unforeseen COVID-19 pandemic, SMEs should think about tactical preparation tools and methods that would have certainly helped them be more versatile when the situation takes an unexpected turn. Identify the clients and end users and involve them during every cycle of tactical planning. Ensure that the enterprise CFT is working toward a single collection of objectives while a different group is working toward a different set of objectives; it authorizes the strategy is falling apart from the track. So, the method has to be interactive engaging all the teams. Have an area built in for calculated positioning, improvisation and recalibration on the quarterly review board urging workers to contribute ideas for improvement.

STRATEGY PLANNING—INTERNAL FACTORS

Strengths, weaknesses, opportunities and threats (SWOT) analysis is a simple yet valuable framework for analyzing. SME management groups need to stick to the foundation; they need to start with an analysis of their present technology and people capacities, prior to setting expectations. All-natural view of the organization is required for starting with the industrial transition. Strengths and weaknesses are interior elements, while opportunities and threats are exterior elements. Typically, it must focus on what is occurring presently versus what can occur in the future. Industries are running in a very profitable market in an ever expanding global industrial economy. So, prioritize threats as well as possibilities as a driver for recalibrating service goals for the future roadmap probably the next one-, four- and seven-year perspectives. Use SWOT evaluation as part of the risk management process to check out and carry out techniques in more balanced and extensive way.

Need to find who are the Enablers, Engagers, Enhancers and have answers for the five W's and one H as

1. *Who is going to be the users? Who is going to be get benefited?*
2. *What are the factors going to add value to the Business?*

3. Where the information comes from and where it goes into?
4. Why it is required?
5. How is going to add value?
6. When it is going to give ROI and ROV?

Use five Whys technique in the analyze phase to do the gap analysis. Followed by Planning, Designing, Implementing, Validating and Rolling out.

The SWOT evaluation process is a brainstorming tool for CFT to discuss various perspectives on the scenario available. Start to craft a method that identifies the competitors to contend successfully in the market. As soon as the four aspects are filled in business executive planner figuring out exactly how each force can be leveraged into opportunities and analyze what weaknesses need to be fixed so that they do not impact the business results considerably. It is an extensive technique for recognizing not just the weak points and hazards of an activity plan, but also the strengths and possibilities it makes possible. When accomplishing the analysis, be sensible as well as strenuous. Constant business analysis and tactical preparation are the most effective means to monitor growth, strengths and weak points. As the claim goes, chance favors the ready mind via the SWOT evaluation; SMEs will certainly be better prepared. With these objectives and actions in hand, SMEs will certainly progress in the direction of completing a strategic plan. Make use of the basic and reliable lean planning for executing the strategic plan. Lean planning continuously fine-tunes and modifies the approach while gauging enterprise's development toward attaining the goals. It is a very easy means to derive and document the strategy, tactics, baseline and forecasts. It is everything about results besides handling an organization's improvement.

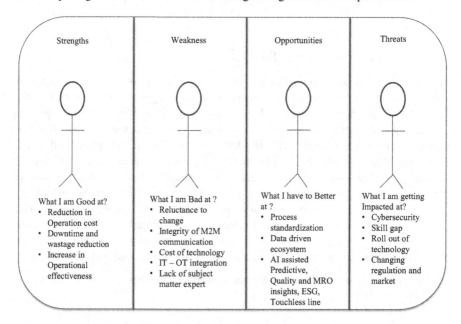

FIGURE 7.2   SWOT analysis of industrial transformation.

STRATEGY PLANNING—EXTERNAL FACTORS

The political, economic, social, technological, legal and environmental (PESTLE) factor evaluation is utilized for macro-environmental scanning and includes the components of PESTLE elements that could have a straight and durable impact on the enterprise. It provides a bird's-eye view of the whole setting from several angles that wish to check and maintain while contemplating a particular business strategy. It serves in all industrial sectors at the strategic, department and examines the existing and future markets. It also provides a vital input for planning, advertising, business and organization transformation, NPD/NPI and so on. Enterprises need to be able to respond to the present and the future legislation and readjust their marketing plan appropriately. It transforms the whole earnings' generation framework such as taxes, trades and fiscal plans, a federal government might impose around, and, it might influence the enterprise's core procedures. Changes and variations to the rising cost of living, international direct investment, stock markets merged as macro- and micro-economical elements will certainly influence the firm's purchasing power, product pricing, market supply demand and will have resonating long-term results. Enterprises need to study health and safety that relates to the social and group trends of the society, which is crucial in identifying the consumer purchasing habits.

Technological transformation opens new doors to technical fads and generates possibilities for many businesses. Both SMEs and OEMs tend to focus heavily on the impact of the emerging new innovations. It plays a vital role in the development of the industrial sectors beyond advancement in the industrial economy. It handles almost everything namely computer numerical control (CNC) machines, PLC, SCADA, industrial robots cyber-security smart production equipment AI cobots and human intelligence. Modern technology is utilized to market, deliver and service end items. It helps in assessing rival information besides comprehending marketing factors, complications and the areas to improve. Enterprises need to comply with customer laws, accessibility to materials, sources, imports and exports, safety criteria, labor regulations and need to understand what is legal and what is not legal in order to trade effectively. Environmental aspects issue the ecological impacts on industry deal with a scenario such as pandemics, global warming and sustainable sources. This is a must have requirement and this is vital because of the increasing depletion

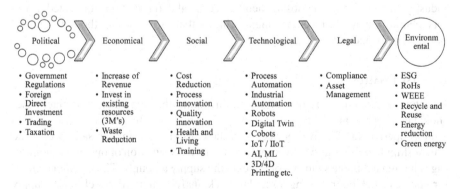

FIGURE 7.3 PESTLE analysis of industrial transformation.

of resources and contamination. It is important to run an industry as a moral and sustainable enterprise, leaving less carbon footprint targets as set by government organizations.

The SWOT analysis focuses on an enterprise's inner strengths and weak points, while PESTLE analysis focuses on the outside variables. Making use of both approaches together creates a thorough assessment for the industrial transformation and offers valuable insights into the enterprise and its standing against the rivals. This leads to another pertinent question when should SWOT analysis as well as PESTLE analysis be utilized? SWOT analysis is utilized to satisfy efficiency requirements preparing for a considerable business transformation to enhance business processes, whereas PESTLE analysis is utilized to understand exactly how the enterprises choose the suit of travel within or toward industrial transformation affecting aspects of the outside world. Development in the industrial sectors is moving at a lightning speed, and NPD and NPI are transforming at an extraordinary rate. Falling behind competitors is a setback for the production venture around the world. SWOT and PESTLE are heavyweight strategic tools that are vital for getting insights, taking advantage of opportunities and reducing the threats of industrial transition. A thorough study will bring big benefits, and the SMEs and OEMs utilize both the tools wherever required for the successful industrial transformation journey.

BUSINESS USE CASE

Manufacturing enterprise leaders are faced with a frustrating number of choices for process automation, process improvement and customer interaction. Bound to combine many self-controls to best offer the customer and realize meaningful organization impacts. Industrial transformation is a journey; to get to the destination, the correct roadmap is essential to drive the transformation in a reliable method. The roadmap starts with an evaluation of the existing culture, skills, structure, capacity, process, jobs, technology, work centers, equipment and innovation maturation and moves on to an interpretation of a future vision, goal and implementation plan. Success relies on how the process transformation is implemented and ensuring that people are ready to accept it and move forward with the trends in technological advancement. Consider a business use case of inventory management in a process industry. Where one of the most difficult and crucial duties that the connected inventory planning plays in inventory monitoring is that of promoting the balancing of inward supply and outward supply.

BUSINESS CHALLENGE

Inventory management is one of the main points the manufacturing enterprises should consider as an obstacle for building their business. Inventory management is vital to operate an effective service, with the consumer's complete satisfaction and reliant shipment times as well as supply control. A key function of inventory monitoring is having a reliable snapshot of the existing supply amounts. The inventory monitor updates stock counts using a regular stock analysis and utilizes effective supply systems to maintain real-time records of stock levels, ensuring that the enterprise supply chain team comprehends its supply placement is perpetually precise. To properly manage the flows in the supply chain, firms have to manage upstream supplier exchanges and downstream customer needs. Manufacturing enterprises dealt with smart manufacturing need to hold an inventory of basic materials, extra components and finished goods in the future smart connected environment.

Precondition

Traditional inventory management has actually been the primary trend in the inventory monitoring space, supplying much better projections showing past sales fads, real-time monitoring and integration with other modern technologies. Traditional stock monitoring helped in decreasing the time invested on manual supply takes as well as uses more accurate data. As technology for services is swiftly enhancing, the use of the future smart connected inventory management will work along with AI, ML, IIoT, cobots and ESG along with human intelligence and will become the norm.

Approach

Data are collected dynamically at each of the process steps. Innovation that makes it possible for future smart inventory management systems comprised of MES, industrial control system (ICS), MRP, radio-frequency identification (RFID) tag, RFID antenna, RFID reader, along with Industry 4.0 and Industry 5.0 technological innovations.

Steps followed are as follows:

- Re-engineering the processes
- Keep track of the vehicle dynamically
- Process traces of the raw material
- Use through opening load cells or pancake load cells for weighing
- RFID along with the industrial technological innovation for tracking.

The MES is integrated with vehicle tracking system (VTS) that is responsible for truck movement from the entrance to the exit. Integrate the information flow from MES

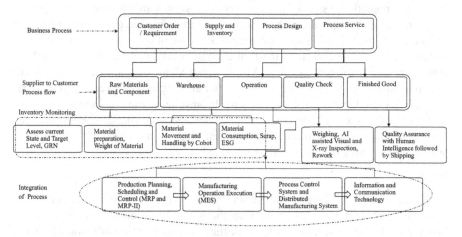

FIGURE 7.4 Process transformation-mapped framework.

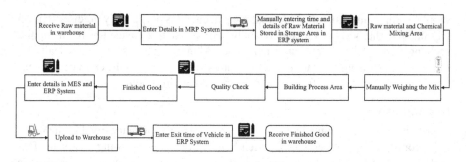

FIGURE 7.5   Before—traditional inventory monitoring system.

along with MRP and VTS. The system captures actual net weighing of all the raw material receipts before generating goods receipt note (GRN). GRN generation and inventory update are mistake-proofed by automating the process. MES helps the shop floor personnel with the display of production plan, recipe selection and the number of final products to be produced, while MRP ensures the issue of raw material as per the plan issued by planning department. MES terminals have RFID reader installed at the packer's stacker. The load cells installed help to capture weight of each batch tagged with storage location. MES will update MRP with the inventory and production information on a periodic basis. MES verifies material with the current running recipe and alerts the user if wrong material is inputted; it ensures process validation and performance of the smart connected inventory system. The connected digital environment will track the actual collective weight of all the input material fed as well as establish traceability of the material till the final finished good through suppliers.

The future smart inventory records the dynamic supply in real net weight; besides, ranges of resources are tape-recorded and updated in actual time, and also, the signals are pushed on to a mobile or over an e-mail to the supervisors, preparation employees and storage personnel. KPIs and reports are given on a dynamic IIoT control board as well as IIoT platform for the whole enterprise with live drill down to the plant level. The future upgraded system connects every possession within a manufacturing organization; this gives exposure to the basic raw material, finished goods, job in progress in addition to their location and problem. Making use of the framework and the connected technology in real time, the future smart stock surveillance system utilized interconnected intelligent systems, AI and cobot along with human touch making handling supply smooth. It produces a smarter and aggressive inventory system can be quickly shared and accessed in real time by anyone, anywhere. In turn, removing hands-on processes, utilizing human intelligence and achieving conformity and getting rid of wastes—waiting time, raw material, manual movement of raw materials, storing and stocking of raw materials, waiting between process steps.

Result

The future smart IIoT-based inventory monitoring and property tracking system use consistent presence right into the stock by offering real-time details brought by RFID tags. It assists to track the accurate location of resources, work in progress and finished products. As an outcome, manufacturers can stabilize

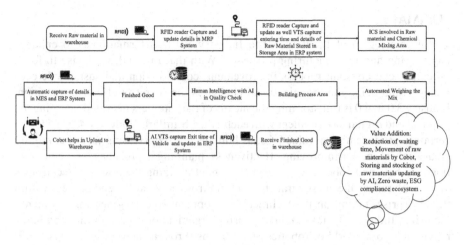

FIGURE 7.6 After—the future smart connected inventory monitoring environment.

the amount of on-hand inventory, boost the usage of equipment, reduce lead time and, therefore, stay clear of concealed costs bound to the much less reliable hand-operated techniques with zero wastage of raw material forming an ESG compliance ecosystem.

WORKPLACE OF THE FUTURE

Technical development helped in the form of employee help systems originated from industrial transformation and can make jobs in the manufacturing sector available for physically challenged persons with the evolution of assistive technology by producing new better systems and changing the workplace of the future. An additional need is the development of discovering systems that enable university graduates to upskill themselves in the transformational modern technologies closing the missing out on gap at a regular time as that of the evolving work environment. Industrial improvement is transforming staff members interact with each other and with the next-generation tools.

> AI have actually progressed to end up being the foundation of several smart connected products and services of the industrial sectors. AI together with Cobots, AI with generative design, AI with self-assistance etc. offers help to physically challenged individuals in work potential as well the opportunity to take part in upgraded industries.
>
> *(Soni et al., 2020)*

Most of the manufacturing enterprises beginning to stay on the top of progressing health and safety regulations that can change instantaneously. Without digital devices, employees are not outfitted to work effectively in the manufacturing facilities of the future. As newer technology disruptions alter the technology landscape, they make way for less costly for open resource innovations that are readily available to SMEs looking to reduce expenses and enhance business transformation processes.

SUMMARY

The ultimate goal of Industry 4.0 to Industry 5.0 transformation is to enhance engineering and manufacturing processes. With that in mind, it is wise to focus on end-to-end constant process improvement, collaboration and sustainable eco-system. Embracing the organizational change is the secret to effective transition. Industry 3.0 to 4.0 is transforming the means of the suppliers operated, merging the physical and digital environment with each other. Similarly, Industry 4.0 to 5.0 will certainly produce higher value jobs, as humans are repossessing jobs that require creative thinking for improving effectiveness, planning approaches for a combination of robots and cobots. It is everything about utilizing the power of electronics along with renewable energy transformation throughout end-to-end process within the enterprise, taking on the technical development from engineering via manufacturing. Industry 5.0 takes manufacturing organizations through the brand-new remarkable joint world of robotics and humans through Internet of Things (IoT)/IIoT and various other modern technological innovations to obtain the work made with accuracy and precision with minimum wastefulness and almost no mistakes. SMEs, part suppliers, OEMs will certainly have the ability to appreciate and enjoying benefits of these technologies with a preliminary financial investment as nothing comes as a complement.

When computers were presented in Industry 3.0, it was turbulent, thanks to the addition of an entirely brand-new innovation. Now, and also into the future as Industry 4.0 unfolds, computer systems are linked and also connect with one another to ultimately make decisions without human involvement. A mix of cyber-physical systems, IoT and the Internet of Solutions make Industry 4.0 possible and the smart manufacturing facility a truth. As a result of the assistance of smart equipment's that maintain getting smarter as they get access to more information, manufacturing facilities will end up being more effective, efficient and less wasteful. Eventually, it is the network of the manufacturers who are digitally connected with one another, develop and share information across the enterprise that results in real power of Industry 4.0. If the current transformation highlights the change of manufacturing facilities into IoT-enabled smart facilities that use cognitive computer, adjoining through cloud computing, Industry 5.0 prepares to focus on the return of human hands along with creative minds into the industrial structure. Industry 4.0 is about the interconnectedness of manufacturers with systems for optimal performance. Industry 5.0 takes such effectiveness, efficiency and human intelligence much better by developing the interaction in between humans with the Industry 4.0 technologies.

Imagine a technology that can provide real-time or instant accessibility to details, along with the computer system power just by idea alone. According to new research study by the United States neuroscientists and nanorobotics scientists, a matrix-style human mind to cloud user interface, that acquaint as Internet of Thoughts network, can be a possibility in the future, i.e., the human mind to cloud interface. By simply utilizing a mix of AI and nanotechnology, researchers stated that human beings will completely have the ability to connect their minds to shadow local area network to gather details from the Web in real time.

Currently, experiencing Industry 4.0 and Industry 5.0, Internet of Thoughts makes sure to bring ourselves on the Internet where it would certainly have the ability to take Industry 6.0 revolution right into our real life. Are you being empowered by the capacity to digitally collaborate with brand-new modern technologies besides the idea of accomplishing new elevations of engineering and manufacturing effectivity that excites as well as motivates? If the response is indeed, then we think alike and move ahead in the industrial transformation journey!

Index

Note: *Italic* page numbers refer to figures.

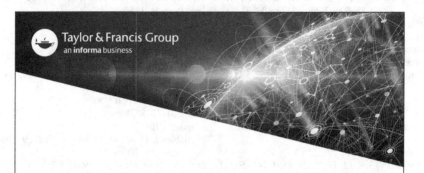

Taylor & Francis Group
an **informa** business

Taylor & Francis eBooks

www.taylorfrancis.com

A single destination for eBooks from Taylor & Francis
with increased functionality and an improved user
experience to meet the needs of our customers.

90,000+ eBooks of award-winning academic content in
Humanities, Social Science, Science, Technology, Engineering,
and Medical written by a global network of editors and authors.

TAYLOR & FRANCIS EBOOKS OFFERS:

A streamlined
experience for
our library
customers

A single point
of discovery
for all of our
eBook content

Improved
search and
discovery of
content at both
book and
chapter level

REQUEST A FREE TRIAL
support@taylorfrancis.com

Routledge
Taylor & Francis Group

CRC Press
Taylor & Francis Group

Printed in the United States
by Baker & Taylor Publisher Services

Printed in the United States
by Baker & Taylor Publisher Services